Lecture Notes
in Economics and
Mathematical Systems
Operations Research, Computer Science, Social Science

Edited by M. Beckmann, Providence, G. Goos, Karlsruhe, and
H. P. Künzi, Zürich

82

R. Saeks

Resolution Space Operators and Systems

Springer-Verlag
Berlin · Heidelberg · New York 1973

Dr. R. Saeks
University of Notre Dame
Dept. of Electrical Engineering
Notre Dame, Ind. 46556/USA

AMS Subject Classifications (1970): 28A45, 46E05, 46G10, 47A99, 47B99, 93A05, 93A10, 93B05, 93B35, 93C05, 93C25, 93C45, 93C50, 93D99.

ISBN 3-540-06155-X Springer-Verlag Berlin · Heidelberg · New York
ISBN 0-387-06155-X Springer-Verlag New York · Heidelberg · Berlin

Offsetdruck: Julius Beltz, Hemsbach/Bergstr.

Preface

If one takes the intuitive point of view that a system is a black box whose inputs and outputs are time functions or time series it is natural to adopt an operator theoretic approach to the study of such systems. Here the black box is modeled by an operator which maps an input time function into an output time function. Such an approach yields a unification of the continuous (time function) and discrete (time series) theories and simultaneously allows one to formulate a single theory which is valid for time-variable distributed and nonlinear systems. Surprisingly, however, the great potential for such an approach has only recently been realized. Early attempts to apply classical operator theory typically having failed when optimal controllers proved to be non-causal, feedback systems unstable or coupling networks non-lossless. Moreover, attempts to circumvent these difficulties by adding causality or stability constraints to the problems failed when it was realized that these time based concepts were undefined and; in fact, undefinable; in the Hilbert and Banach spaces of classical operator theory.

Over the past several years these difficulties have been alleviated with the emergence of a new operator theory wherein time related concepts such as causality, stability and passivity are well defined and readily studied. Although the techniques were developed essentially independently by a number of different authors they are characterized by two consistent threads. First, the time characteristics of a function (or vector) are characterized by a family of truncation operators which pick out the part of the function before and after a given time. Secondly, various operator valued integrals manifest themselves as the primary tool with which the theory is implemented.

The most straightforward approach has been that of Willems, Damborg, Zames and Sandberg who, in the context of their researches into feedback system stability, adopted the point of view of working with a function space together with the family of truncation operators defined by

$$(E^t f)(s) = \begin{cases} f(s) \ ; & s \leq t \\ 0 \ ; & s > t \end{cases}$$

Here $E^t f$ is interpreted as the part of f before time t and $f - E^t f$ is interpreted as the part of f after t. As such, by appropriately associating the operators on the given function space with the family of truncation operators, the various stability and related causality concepts are defined. Closely related to this work is the Fourier analysis of Falb and Freedman who deal with similar function spaces and families of truncation operators. Here the Fourier representation and corresponding theorems are formulated via appropriate weakly convergent operator valued integrals in a manner somewhat similar to that used in a spectral theoretic context.

Somewhat more abstract than the above is the work of Duttweiler and Kailath who deal with a Hilbert space of stochastic processes in an estimation theoretic context. Here, however, because of the predictability of such stochastic processes the truncation operators are replaced by a resolution of the identity wherein $E^t f$ is an estimate of f conditioned on measurements made prior to time t, and the integrals of triangular truncation, originally formulated by the Russian school of spectral theorists, serve as the primary manipulative tool. An alternative approach, also characterized by truncation operators and operator valued measures is due to Zemanian. Here Hilbert and/or Banach space valued distributions are employed for the study of various problems associated with the realizability theory of linear networks.

Finally, we have the resolution space approach of the present monograph wherein abstract Hilbert space techniques are combined with classical spectral theory to obtain a formalism which essentially includes all of the above, from which many of the ideas and techniques have been drawn. The approach is due primarily to Porter and the author whose interests in the subject were derived, respectively, from their research in optimal control and network synthesis. In this theory a resolution of the identity on an abstract Hilbert space defines the required truncation operators, which are manipulated via the Lebesgue and several different types of Cauchy integral defined over appropriate operator valued measures.

In the first chapter the elements of resolution space and the theory of causal operators are formulated. These ideas are applied to the study of the stability, sensitivity and optimization of feedback systems in the second chapter, and to the controllability, observability, stability and regulator problems for dynamical systems in the third chapter. The last chapter deals with the more highly structured concept of a uniform resolution space wherein time-invariant operators, along with their Fourier and Laplace transforms, are developed. Finally, there are four appendices. The first three of these are devoted to a review (without proof) of results from topological group theory, operator valued integration, and spectral theory, respectively, which are needed in the main body of the text, while the representation theory for resolution and uniform resolution spaces is formulated in the fourth appendix.

Although the placement of the applied chapters in the middle of the text may seem to be somewhat unusual, this ordering was chosen so that the three elementary chapters (equals a first graduate course in real analysis) would appear first with the more advanced material (equals harmonic analysis and non-self-adjoint spectral theory) in the last chapter and appendices. As such, the first three chapters may be used for a first or second year graduate course in either engineering or mathematics. This being aided by rather long discussion and problem sections at the end of each chapter. These include problems which range from the simple through difficult to the open. In fact, many generalizations of the theory and related topics are introduced in the problems.

References to results in the monograph are made by a three digit characterization, for instance (2.D.3) denotes the third paragraph in the fourth section of the second chapter. In general theorems, lemmas and equations are unnumbered with references made to the chapter section and paragraph only. Since the paragraphs contain at most one "significant" result, this causes no ambiguity. Similarly, references to books and technical papers are made via a three digit characterization such as (FG-3) which denotes the third paper by author FG. With regard to such references, except for classical theorems, we have adopted the policy of not attributing theorems to specific people though we often give a list of papers wherein results related to a specific theorem appear. This policy having been necessitated by the numerous hands through which most of the results presented have passed in the process of reaching the degree of generality in which they are presented.

Finally, acknowledgements to those who have contributed to this work via their comments and suggestions are in order. To list but a few: R.A. Goldstein, R.J. Leake, N. Levan, S.R. Liberty, R.-W. Liu, P. Masani,

W.A. Porter, M.K. Sain and A.H. Zemanian. Moreover, I would like to express my gratitude to R.M. DeSantis for his careful reading and detailed comments on the manuscript, to P.R. Halmos for recommending inclusion of the manuscript in the Springer-Verlag Lecture Notes series, and to Mrs. Louise Mahoney for her careful typing of the manuscript.

Contents

1. Causality.. 1

A. Resolution Space.. 3

B. Causal Operators.. 8

C. Closure Theorems.. 14

D. The Integrals of Triangular Truncation............................ 22

E. Strictly Causal Operators... 29

F. Operator Decomposition.. 41

G. Problems and Discussion... 50

2. Feedback Systems.. 61

A. Well-Posedness.. 63

B. Stability... 72

C. Sensitivity... 82

D. Optimal Controllers... 89

E. Problems and Discussion... 100

3. Dynamical Systems... 114

A. State Decomposition... 116

B. Controllability, Observability and Stability...................... 132

C. The Regulator Problem... 140

D. Problems and Discussion... 148

4. Time-Invariance... 155

A. Uniform Resolution Space.. 156

B. Spaces of Time-Invariant Operators................................ 160

C. The Fourier Transform... 168

D. The Laplace Transform... 178

E. Problems and Discussion 191

Appendices

A. Topological Groups.. 204

A. Elementary Group Concepts.............................. 205

B. Character Groups.. 206

C. Ordered Groups.. 209

D. Integration on (LCA) Groups............................. 211

E. Differentiation on (LCA) Groups......................... 214

B. Operator Valued Integration............................... 217

A. Operator Valued Measures................................ 219

B. The Lebesgue Integral................................... 222

C. The Cauchy Integrals.................................... 225

D. Integration over Spectral Measures...................... 229

C. Spectral Theory... 232

A. Spectral Theory for Unitary Groups...................... 233

B. Spectral Multiplicity Theory............................ 237

C. Spectral Theory for Contractive Semigroups.............. 240

D. Representation Theory..................................... 244

A. Resolution Space Representation Theory.................. 245

B. Uniform Resolution Space Representation Theory......... 248

.... 258

1. CAUSALITY

In this chapter the basic concepts of resolution space and causal operators which prove to be of fundamental importance throughout the work are introduced. The results presented are due primarily to W. A. Porter (PR-1, PR-2, PR-3, PR-4), R. M. DeSantis (DE-1, DE-2, DE-3) and the author (SA-4) in their present formulation. Similar results have, however, been obtained in the context of other formalisms (WL-3, DA-1, DA-2, DA-3, DU-1, DI-1, ZE-1, ZE-2). We begin with a formulation of the basic definitions and notation for resolution space along with a number of examples which are used throughout the remainder of the work. We then introduce causal operators, which prove to be the natural operators with which one deals in a resolution space setting, and the related concepts of anti-causal and memoryless operators. Various algebraic and topological closure theorem are then formulated for these operator classes with special emphasis being placed on causal invertibility theorems.

Our first encounter with operator valued integrals in our theory is with the integrals of triangular truncation which define (unbounded) projections of the space of all bounded linear operator onto the space of causal operators. Since the causal operators are algebraically defined it is not surprising that these integrals are independent of the operator topology in which they are defined. To the contrary, however, if one changes the bias on the Cauchy integrals used to define the integrals of triangular truncation to obtain the integrals of strict triangular truncation they are topology dependent. As such these latter integrals define various classes of strictly causal operators depending on the operator topology employed. Finally, the problem of decomposing a general operator into a sum or product of operators with specified causality characteristics is studied.

Since the theories of causal and anti-causal operators (strictly causal and strictly anti-causal) are dual, for the sake of brevity we prove only theorems about the causal case with the corresponding theorems for the anti-causal case being stated without proof or skipped completely. Finally, a number of concepts related to causal operators, especially dealing with the causality of nonlinear operators, are discussed in the final section.

A. Resolution Space

1. Definition and Notation

By a (Hilbert) resolution space, we mean a triple, (H,G,E), where H is a Hilbert space, G is an ordered (LCA) topological group and E is a spectral measure defined on the Borel sets of G whose values are projections on H. Typically the group G is fixed throughout any development and hence we suppress it in our notation denoting (H,G,E) simply by (H,E). Since G is ordered E uniquely determines a resolution of the identity via

$$E^t = E(-\infty,t)$$

and, conversely, such a resolution of the identity determines a spectral measure hence the term resolution space. Similarly, we denote by E_t the projection valued function

$$E_t = E(t,\infty) = I - E^t$$

where I is the identity operator, and we denote by H^t and H_t the ranges of E^t and E_t, respectively. Clearly, H^t and H_t are orthogonal complements with H^t (H_t) going to H(0) as t goes to ∞ and to 0(H) as t goes to $-\infty$.

Throughout the remainder of the monograph all operators are assumed to be bounded linear operators mapping a resolution space (H,E) into itself unless specified to the contrary.

Since the spectral measure E and the projection valued functions E^t and E_t play a significant role in much of our theory we tabulate some of their basic properties below.

i) $E(A) = E(A)^*$, $E^t = E^{t*}$, $E_t = E_t{}^*$ for all Borel sets, A, and t in G.

ii) $$E(G) = \underset{t \to \infty}{\text{s-limit}}\, E^t = \underset{t \to -\infty}{\text{s-limit}}\, E_t = I$$

where s-limit denotes the limit in the strong operator topology.

iii) $$E(\emptyset) = \underset{t \to -\infty}{\text{s-limit}}\, E^t = \underset{t \to \infty}{\text{s-limit}}\, E_t = 0$$

iv) $E(A)E(B) = E(B)E(A) = E(A)$, $E^tE^s = E^sE^t = E^t$, and $E_tE_s = E_sE_t = E_s$ if A is contained in B and $t \leq s$.

v) $E(s,t) = E^t - E^s = E_s - E_t = E^tE_s = E_sE^t$ for all $s \leq t$.

vi) $E(A)E(A) = E(A)$, $E^tE^t = E^t$, abd $E_tE_t = E_t$ for all Borel sets A and t in G.

vii) For all Borel sets A and t in G the norm of $E(A)$, E^t and E_t is either one or zero.

2. Examples

The most common example of a resolution space which is used as a model for much of our theory is the space (H,E) where

$$H = L_2(G,K,\mu)$$

is the Hilbert space of functions defined on an ordered (LCA) group, G, taking values in a Hilbert space, K, which are square integrable relative to a Borel measure, μ, and E is the spectral measure defined by multiplication of the characteristic function

$$(E(A)f)(s) = \chi_A(s)f(s)$$

for each Borel set, A. Here, E^t reduces to the family of truncation operators

$$(E^t f)(s) = \begin{cases} f(t) & ; s \leq t \\ 0 & ; s > t \end{cases}$$

and H^t is the set of functions which are zero (a.e.) for $s > t$. This structure includes most of the classical spaces encountered in system theory wherein G is taken to be R or Z, with their natural orderings, $K = R^n$ and μ is the Lebesgue or counting measure on R or Z respectively. Moreover, all of our definitions for classical systems concepts coincide with their classical counterpart in these spaces.

Of course, we may take H to be a subspace of $L_2(G,K,\mu)$ with the same spectral measure to obtain a number of alternative resolution spaces. For instance functions which are nonzero only over a specified time set (i.e., subset of G), such as l_2^+ which may be interpreted as the subset of $L_2(Z,R,m)$, m the counting (=Haar) measure on Z, composed of functions which are zero for $t < 0$. Similarly, L_2^+ may be interpreted as a subspace of $L_2(R,R,m)$. In general, we identify <u>forward half-spaces</u> of $L_2(G,K,m)$ with $L_2(G,K,m^+)$ where m^+ is the restriction of the Haar measure to G^+.

In general, for any function space such as those described above we term the spectral measure defined by multiplication by the characteristic function the <u>usual resolution space structure</u> or the <u>usual spectral measure</u> for the function space and assume that all function spaces are equipted with this spectral measure without explicitly including it in the notation unless specified to the contrary.

Resolution space is primarily a setting for the study of time related, hence infinite dimensional, problems. The concept is, however, well defined, though somewhat degenerate, in the finite dimensional case. In fact, if (H,E) is a resolution space with H finite dimensional we are assured that E^t is constant except for a finite number of "jumps" at t_1, t_2, ..., t_k; $k \leq n = \dim(H)$ (since H^t is increasing from 0 to H and can do this at most n times before reaching H). Now if one lets $K^1 = H^1$, $K^2 = H^2/H^1$ and $K^i = H^i/H^{i-1}$, i = 1, 2, ..., k, we may identify (H,E) with the space of functions defined on the set t_i; i = 1, 2, ..., k with $f(t_i)$ in K^i, together with the spectral measure defined by multiplication by the characteristic function. Thus every finite dimensional resolution space is of this highly degenerate (and thus uninteresting) form.

Intuitively, if x is in H we interpret $E^t x$ as that part of x before t and $E_t x$ as that part of x after t. Unlike the case of a deterministic signal wherein the concepts of past and future are unambiguous in the case of a Hilbert space of stochastic processes several alternative points of view are possible, each of which yields a distinct but well defined, resolution space with the Hilbert space, H, of stochastic processes remaining fixed but with the spectral measure, E, dependent on the concepts of past and future which one desires to employ. Of course, we may let E be the usual spectral measure defined by multiplication by the characteristic function in which case the part of the stochastic process before t has its usual interpretation. On the other hand since the portion of a stochastic process before t contains statistical information about that part of the process appearing after t we may desire to define the past as all information about the signal (whether before or after t) which can be obtained from measurements before t. One way of achieving this is to define $E^t f$ via

$$(E^t f)(s,\omega) = \begin{cases} f(s,\omega) & s \leq t \\ \underline{f}(s,\omega) & s > t \end{cases}$$

for each t,s and event, ω. Here $\underline{f}$ is taken to be the best linear estimate of f after t based on measurements before t. Of course, with this definition of the past the future, $E_t f$, is the error between the best estimate of the future and the actual future. Assuming that the Hilbert space of stochastic processes is so defined as to allow such best estimates E^t is well defined and a resolution of the identity. As such, it uniquely determines a spectral measure and a resolution space of stochastic processes (DU-1).

3. Product and Sub-Resolution Spaces

If one fixes G the concepts of product and sub-resolution space may be defined in their natural manner. In particular, given resolution spaces (H,E) and $(\underline{H},\underline{E})$ we define their product resolution space via

$$(H,E)\otimes(\underline{H},\underline{E}) = (H\otimes\underline{H},E\otimes\underline{E})$$

where $H\otimes\underline{H}$ is the usual direct product of Hilbert spaces and $(E\otimes\underline{E})(A)$ is the projection onto the direct product of the images of E(A) and $\underline{E}(A)$. Similarly, if K is a subspace of H such that E(A) maps K into itself for all A then (K,E_K) is a sub-resolution space where E_K is the restriction of E to K.

B. Causal Operators

1. Causality Definitions

We say that a linear bounded operator T on a resolution space (H,E) is causal if for any x, y in H and t in G such that $E^t x = E^t y$ then $E^t Tx = E^t Ty$. In essence causality corresponds to the statement that T cannot predict the future (as per the definition of future implied by the resolution space structure) by assuring that the operator cannot distinguish, before t, between inputs which are identical before t.

In the case of an operator mapping a resolution space (H,E) into a resolution space $(\underline{H},\underline{E})$, both over the same group, we say it is causal if for any x,y in H and t in G such that $E^t x = E^t y$ then $\underline{E}^t Tx = \underline{E}^t Ty$. Clearly, all of the causality properties in this case are the same as for the case of an operator mapping a space into itself hence for notational brevity in both our definitions and theorems we assume identical range and domain spaces, though we employ the results in the general case when required.

A number of alternative, but equivalent, definitions for the concept of causality have appeared in the literature (WI-1). Some of these are characterized by the following theorem.

Theorem: For a linear operator T the following are equivalent.

i) T is causal.

ii) $E^t T = E^t T E^t$ for all t in G.

iii) $||E^t Tx|| \leq ||TE^t x||$ for all x in H and t in G.

iv) If $E^t x = 0$ then $E^t Tx = 0$ for all x in H and t in G.

v) H_t is an invariant subspace of T (i.e., $T(H_t) \subset H_t$) for all t.

vi) $TE_t = E_t T E_t$ for all t.

vii) $E^t T E_t = 0$ for all t.

Proof: <u>i) => ii).</u> Since E^t is a projection $E^t(E^tx) = E^tx$ hence causality implies that $E^tTE^tx = E^tTx$ for all t and x, hence ii).

<u>ii) = iii).</u> If $E^tTx = E^tTE^tx$ then

$$||E^tTx||^2 = ||E^tTE^tx||^2 \leq ||E^tTE^tx||^2 + ||E_tTE^tx||^2$$
$$= ||TE^tx||^2$$

implying iii). Note that the last equality results from the fact that E^tTE^tx and E_tTE^tx are orthogonal.

<u>iii) => iv).</u> If $E^tx = 0$ then linearity assures that $TE^tx = 0$ hence so is its norm whence iii) implies that $||E^tTx|| = 0$ hence so is E^tTx itself verifying iv).

<u>iv) => v).</u> If $x \in H_t$ then $E^tx = 0$ hence iv) implies that $E^tTx = 0$ hence $Tx \in H_t$ verifying that H_t is indeed invariant.

<u>v) => vi).</u> If H_t is invariant then for any x, $TE_tx \in H_t$. Thus since E_t is the projection onto H_t, $E_tTE_tx = TE_tx$ implying vi).

<u>vi) => vii).</u> If $E_tTE_t = TE_t$ then

$$E^tTE_t = E^tE_tTE_t = 0$$

since $E^tE_t = E^t(I-E^t) = E^t - E^t = 0$.

<u>vii) => i).</u> If $E^tx = E^ty$ then $E^t(x-y) = 0$ hence vii) implies that

$$0 = E^tTE_t(x-y) = E^tT(x-y)$$

or equivalently since T is linear, $E^tTx = E^tTy$ verifying i), and completing the proof of the theorem.

2. Examples

Consider the resolution space $L_2(G,K,m)$, m the Haar measure on G, with its usual resolution structure, and an operator, T, mapping this space into itself such that y = Tu is defined by

$$y(t) = \int_G \underline{T}(t,q)u(q)dm(q)$$

Here T is assured to be well defined if $\underline{T}$ is square integrable as a function of two variables (relative to the Haar measure on GxG) and may in fact be more generally defined. Now T is causal relative to the usual resolution structure if and only if $\underline{T}(t,q) = 0$ for $t < q$. To see this assume that $\underline{T}(t,q) = 0$ for $t < q$ and that u is in H_s then $u(q) = 0$ a.e. for $q \leq s$ hence for $t \leq s$ we have

$$y(t) = \int_G \underline{T}(t,q)u(q)dm(q)$$

$$= \int_{(-\infty,t]} \underline{T}(t,q)u(q)dm(q) + \int_{(t,\infty)} \underline{T}(t,q)u(q)dm(q)$$

Now for the first integral $q \leq t \leq s$ hence u(q) and the integral is zero whereas for the second integral $q > t$ hence $\underline{T}(t,q)$ is zero and so is the integral. We thus have $y(t) = 0$ for $t \leq s$ showing that y is in H_s and that T is causal. Since a similar argument yields the converse relationship our contention is verified.

For finite dimensional resolution space we have already indicated that any such space may be interpreted as a function space defined on an ordered set of elements $t_1, t_2, \ldots, t_k$ $k \leq n = \dim(H)$ with $f(t_i)$ in K^i a family of finite dimensional vector spaces such that $\sum_{i=1}^{k} \dim(K^i) = n$ (1.A.2). As such, if one identifies an element of H with a partitioned n-vector $x = \text{col}(f(t_1), f(t_2), \ldots, f(t_k))$ then any linear operator on H may be represented as a partitioned n by n matrix, and the operator will be causal if and only if the matrix is (block) lower triangular. This follows from essentially the same argument which was used for the case of the convolution operator with the integral replaced by the summation of matrix multiplication.

As a final example we consider a Toepliz operator on $l_2^+ = L_2(Z,R,m^+)$ represented by a semi-infinite matrix with $\underline{T}_{i,j} = \overline{T}_{i-j}$ where $\overline{T}$ is a specified sequence. Now upon identifying the semi-invinite matrix multiplication with a convolution the Toeplitz operator is found to be causel if and only if $\overline{T}_k = 0$ for $k < 0$ or equivalently the Toeplitz matrix $\underline{T}$ is lower triangular since $\underline{T}_{i,j} = \overline{T}_{i-j} = 0$ if $i > j$.

3. Anti-Causal Operators

We say that a linear bounded operator, T, on a resolution space (H,E) is anti-causal if for any x,y in H and t in G such that $E_t x = E_t y$ then $E_t Tx = E_t Ty$. In essence anti-causality implies that an operator cannot remember the past, though it may be able to predict the future. The concept of anti-causality is thus dual to that of causality and all of our theorems in one case naturally translate into the other. As such, most of our theorems on anti-causal operators are stated without proof since the arguments required are essentially the same as for the corresponding theorems on causal operators.

Similarly, our examples of causal operators become examples of anti-causal operators once the various inequalities have been reversed and the lower triangular matrices replaced with upper triangular matrices.

The dual of our characterization theorem for causal operators is:

Theorem: For a linear operator, T, the following are equivalent.

i) T is anti-causal.

ii) $E_t T = E_t T E_t$ for all t in G.

iii) $||E_t Tx|| \leq ||TE_t x||$ for all x in H and t in G.

iv) If $E_t x = 0$ then $E_t Tx = 0$ for all x in H and t in G.

v) H^t is an invariant subspace of t (i.e., $T(H^t) \subset H^t$) for all t.

vi) $TE^t = E^t TE^t$ for all t.

vii) $E_t TE^t = 0$ for all t.

4. Memoryless Operators

Since an anti-causal operator cannot remember the past and a causal operator cannot remember the future it is natural to term an operator which is both causal and anti-causal memoryless. Such operators "remember" only the present. In particular, for a convolution operator to be causal we must have $\underline{T}(t,q) = 0$ for $t < q$ and for it to be anti-causal we require $\underline{T}(t,q) = 0$ for $q < t$ hence it is memoryless if and only if $T(t,q) = 0$ for $t \neq q$. As such the convolution (with distributional weighting function) reduces to a multiplication $y(t) = \overline{T}(t)u(t)$. Similarly, the matrix representations of a memoryless operator are both upper and lower triangular, hence diagonal.

The characterization theorems for causal and anti-causal operators combine to yield the following characterization for memoryless operators.

Theorem: For a linear bounded operator, T, the following are equivalent.

i) T is memoryless.

ii) H^t and H_t are invariant subspaces of T for all t.

iii) $TE(A) = E(A)T$ for all Borel sets A of G.

iv) $\hat{U}^\gamma T = T\hat{U}^\gamma$ for all γ in G_T where $\hat{U}^\gamma = L\int_G (\gamma,t)_T dE(t)$ is the group of unitary operators associated with E via Stone's theorem.

Proof: The equivalence of i) and ii) follows immediately from the characterization theorems for causal and anti-causal operators whereas the equivalence of iii) and iv) follows from Stone's theorem (C.A.2) hence it suffices to show that i) and iii) are equivalent. Now if T is memoryless then it is both causal and anti-causal hence

$$E^tT = E^tTE^t = TE^t$$

but the commutativity of T with the resolution of the identity defined by E is equivalent to its commutativity with E; hence i) implies iii). Conversely, if E(A)T = TE(A) for all A then since E^t is a projection

$$E^tT = E^tE^tT = E^tTE^t$$

and

$$TE^t = TE^tE^t = E^tTE^t$$

showing that T is both causal and anti-causal, hence memoryless, and the proof of the theorem is complete.

C. Closure Theorems

1. Operator Algebras

The various classes of operator which we have thus far defined are mathematically well behaved as per the following theorem.

Theorem: The set of causal (anti-causal, memoryless) bounded linear operators form a Banach Algebra with identity which is closed in the strong operator topology of the algebra of all bounded linear operators.

Proof: The proofs for the causal and anti-causal case are dual while the result for the memoryless case follows from that for the causal and anti-causal case since the intersection of two Banach Algebras is also a Banach Algebra; hence, it suffices to prove the theorem for the case of causal operators. If T and S are causal then

$$E^tTS = E^tTE^tS = E^tTE^tSE^t = E^tTSE^t$$

hence TS is also causal. Similarly,

$$E^t(T+S) = E^tT + E^tS = E^tTE^t + E^tSE^t = E^t(T+S)E^t$$

showing that T+S is causal. We also have $E^tI = E^tE^tI = E^tIE^t$ showing that the identity is causal. Finally, if T_i is a sequence of causal operators converging strongly to T, i.e.,

$$\lim_{i \to \infty} T_ix = Tx$$

for all x then since E^t is bounded

$$E^tTx = E^t[\lim_{i \to \infty} T_ix] = \lim_{i \to \infty} E^tT_ix = \lim_{i \to \infty} E^tT_iE^tx =$$

$$= E^t[\lim_{i \to \infty} T_i E^t x] = E^t T E^t x$$

showing that T is causal and thus completing the proof of the theorem.

Note that since the space of causal operators is closed in the strong operator topology it is also closed in the uniform topology since uniform convergence implies strong convergence.

2. Adjoints

In general, the adjoint of a causal operator is not causal. In fact:

Theorem: The adjoint of every causal operator is anti-causal and conversely.

Proof: If $E^t T E_t = 0$ then since $0^* = 0$ we have

$$0 = 0^* = (E^t T E_t)^* = E_t T^* E^t$$

since E_t and E^t are self adjoint. T^* is thus anti-causal. Now a similar argument beginning with an anti-causal T proves the converse and the proof of the theorem is complete.

Note that since a memoryless operator is both causal and anti-causal so is its adjoint; hence, the adjoint of a memoryless operator is memoryless. We also note that if a self adjoint operator is causal it must also be anti-causal since it is equal to its adjoint. Thus the only causal (anti-causal) self adjoint operators are memoryless and similarly for skew adjoint operators ($T^* = -T$). Along the same lines a triangular matrix argument will assure that the inverse of a causal finite dimensional operator is causal. As such, any causal finite dimensional unitary operator is also memoryless (since its causal inverse is equal to its anti-causal adjoint; hence, the adjoint and therefore the operator itself must be memoryless). Fortunately, this degeneracy does not take place in the infinite dimensional case wherein causal unitary and isometric operators play a fundamental role. This phenomenon thus illustrating the essential infinite dimensionality of our theory.

3. Causal Invertibility

Unlike the case of the adjoint the problem of characterizing the inverses of causal operators is much more difficult. Of course, in the finite dimensional case triangular matrix arguments will suffice to assure that the inverse of a causal operator is causal while the commutativity condition characterizing the memoryless operators is preserved under inversion - hence the inverse of a memoryless operator, when it exists is also memoryless. In the case of an arbitrary causal (or anti-causal) operator we have the following theorem (DA-1).

Theorem: Let T be an arbitrary invertible operator. Then T^{-1} is causal if and only if whenever $E^t x \neq 0$ then $E^t Tx \neq 0$ and T^{-1} is anti-causal if and only if whenever $E_t x \neq 0$ then $E_t Tx \neq 0$.

Proof: The proofs for the causal and anti-causal case are dual and we only consider the causal case. For T^{-1} to be causal we require for any y such that $E^t y = 0$ that $E^t T^{-1} y = 0$ or equivalently, upon invoking the contrapositive form of the above statement we require that if $E^t T^{-1} y \neq 0$ then $E^t y \neq 0$. Now since T is invertible every x is of the form $x = T^{-1} y$ for some y hence the above statement is equivalent to the requirement that $E^t x \neq 0$ implies that $E^t Tx \neq 0$ which was to be shown.

An alternative condition for the existence of a causal inverse which presupposes that the given operator is causal and simultaneously verifies the existence of a bounded inverse and its causality is the following.

<u>Theorem</u>: Let T be a causal bounded operator. Then T has a causal bounded inverse if and only if there exists an $\varepsilon > 0$ such that

$$||E^t Tx||^2 + ||E_t T^*x||^2 \geq \varepsilon \, ||x||^2$$

for all t and x.

Proof: If the above condition is satisfied then by letting t go to infinity we have

$$||Tx||^2 \geq \varepsilon \, ||x||^2$$

for all x in H and by letting t go to minus infinity we have

$$||T^*x||^2 \geq \varepsilon \, ||x||^2$$

for all x in H. Now if $x \neq 0$, $||x||^2 > 0$ and we have

$$||Tx||^2 \geq \varepsilon \, ||x||^2 > 0$$

and

$$||T^*x||^2 \geq \varepsilon \, ||x||^2 > 0$$

Hence $Tx \neq 0$ and $T^*x \neq 0$. T and T* are thus both one-to-one and since the adjoint of a one-to-one operator has dense range T must have dense range. Now for any y in H let y_i be a sequence of vectors in the range of T converging to y, and let x_i be the preimages of the y_i under T (i.e., $y_i = Tx_i$). For any j and k we have

$$||y_j - y_k||^2 = ||Tx_j - Tx_k||^2 = ||T(x_j - x_k)||^2 \geq \varepsilon \, ||x_j - x_k||^2$$

and since the y_i converge to y they form a Cauchy sequence; hence, the x_i also form a Cauchy sequence and thus converge to some x in H. Now since T is bounded it is continuous and thus

$$y = \underset{i \to \infty}{\text{limit}}\, y_i = \underset{i \to \infty}{\text{limit}}\, Tx_i = T[\underset{i \to \infty}{\text{limit}}\, x_i] = Tx$$

showing that y is in the range of T and thus that T is onto. Since T is one-to-one and onto its inverse exists and for any z in H we have

$$||z||^2 = ||T(T^{-1}z)||^2 \geq \varepsilon\, ||T^{-1}z||^2$$

hence T^{-1} is bounded with norm less than or eq,al to $1/\varepsilon$. Finally, upon letting $x = E^tT^{-1}z$ in the original equality of the hypotheses we have

$$||E^tTE^tT^{-1}z||^2 + ||E_tT^*E^tT^{-1}z||^2 \geq \varepsilon\, ||E^tT^{-1}z||^2$$

Now since T is causal T^* is anti-causal; hence, $E_tT^*E^t = 0$ while the causality yields

$$E^tTE^tT^{-1}z = E^tTT^{-1}x = E^tz$$

which upon substitution into the above equation yields

$$||E^tz||^2 \geq \varepsilon\, ||E^tT^{-1}z||^2$$

Hence if $E^tz = 0$, $E^tT^{-1}z$ must also be zero verifying that T^{-1} is causal.

Conversely, if T^{-1} exists, is causal and has norm M, then for any y our norm condition for causality (1-B.1) yields

$$||E^tT^{-1}y||^2 \leq ||T^{-1}E^ty||^2 \leq M^2||E^ty||^2$$

Similarly, if T^{-1} is causal T^{*-1} is anti-causal and our norm condition for anti-causality(1.B.1) yields

$$||E_t T^{*-1} z||^2 \leq || T^{*-1} E_t z||^2 \leq M^2 ||E_t z||^2$$

for any z. Letting z = T*x and y = Tx thus yields

$$||E^t x||^2 = ||E^t T^{-1} Tx||^2 \leq M^2 ||E^t Tx||^2$$

and

$$||E_t x||^2 = ||E_t T^{*-1} T^* x||^2 \leq M^2 ||E_t T^* x||^2$$

Finally, upon letting $\varepsilon = 1/M^2$ and adding these two equations we have

$$||E^t Tx||^2 + ||E_t T^* x||^2 \geq \varepsilon \, ||E^t x||^2 + \varepsilon \, ||E_t x||^2 = \varepsilon \, ||x||^2$$

where the last equality results from the orthogonality of $E_t x$ and $E^t x$. This is the required inequality and the proof is thus complete.

4. Sufficient Conditions for Causal Invertibility

Although the above conditions for causal invertibility are necessary and sufficient they are not readily applicable. Rather we desire a test based wholly on the Hilbert space characteristics of the operator, and possibly, an a-priori assumption that the original operator is causal.

One such causal invertibility theorem (SA-4, PR-4) requires that an operator be definite if its inverse is to be causal. That is, an operator T is definite if $\langle x, Tx \rangle \neq 0$ for all $x \neq 0$ in H. The definite operators, which include the positive and negative definite operators and operators with positive and negative definite hermitian parts, are characterized by the following Lemma.

Lemma: If T is a definite operator then

i) T^{-L} exists (i.e., T has a left inverse).

ii) T^{-1} exists if and only if T^{-L} is definite.

Proof: i). If T^{-L} does not exist T is not one-to-one hence there must exist an $x \neq 0$ such that $Tx = 0$ in which case $\langle x,Tx\rangle = \langle x,0\rangle = 0$ showing that T is not definite and verifying i).

ii). If T^{-L} is definite i) implies that $(T^{-L})^{-L}$ exists hence T^{-L} is one-to-one implying that T is onto and thus, since it is one-to-one by i), invertible. Conversely, if T^{-1} exists but is not definite then there exists $x \neq 0$ such that $\langle x,T^{-1}x\rangle = 0$ and we may let $y = T^{-1}x \neq 0$ (since T^{-1} is one-to-one) yielding

$$\langle y,Ty\rangle = \overline{\langle Ty,y\rangle} = \overline{\langle TT^{-1}x,T^{-1}x\rangle} = \overline{\langle x,T^{-1}x\rangle} = 0$$

which contradicts the definiteness of T and completes the proof.

With the aid of the preceeding lemma we may now prove:

Theorem: Let T be a causal invertible operator. Then T^{-1} is causal if T^n is definite for some $n \geq 1$.

Proof: The case for $n > 1$ follows trivially from the case for $n = 1$ hence we will prove that case first. Let x be in H_t and define y by

$$y = (I-TE_tT^{-1})x$$

for any t in G. Now y is in H_t since x is in H_t and T is causal (hence T takes the range of E_tT^{-1} into H_t) and we have

$$T^{-1}y = T^{-1}(I-TE_tT^{-1})x = (T^{-1}-E_tT^{-1})x$$

$$= (I-E_t)T^{-1}x = E^tT^{-1}x \in H^t$$

$T^{-1}y$ is thus orthogonal to y and we have $\langle y,T^{-1}y\rangle = 0$ whence by the lemma

$y = 0$ and $x = TE_tT^{-1}x$ hence upon multiplying through by T^{-1} we have

$$T^{-1}x = T^{-1}TE_tT^{-1}x = E_tT^{-1}x \in H_t$$

showing that T^{-1} is causal as required. Finally, if T^n is definite for some $n > 1$ then the theorem for the case $n = 1$ applies to the operator (T^n) which is causal and invertible; hence $(T^n)^{-1}$ is causal and

$$T^{-1} = (T^n)^{-1}T^{n-1}$$

is causal since it is the product of causal operators. The proof of the theorem is therefore complete.

D. The Integrals of Triangular Truncation

1. Lower Triangularity

In all of our examples wherein we have characterized the causal operators in terms of the properties of some representation it has always been in the form of some type of triangularity concept. That is, either a triangular matrix representation or a convolutional weighting function whose support is contained in a triangular region. These triangularity concepts can be extended to the case of an abstract operator via the integrals of triangular truncation (GO-1, GO-3, SA-4, DE-1, BR-1) characterized in the following lemma.

Lemma: For any bounded operator T

$$RC\int_G dE(t)TE^t = LC\int_G E_t T dE(t)$$

(i.e., one integral exists if and only if the other exists and their values coincide when they exist).

Proof: For any finite partition of G $-\infty = t_0 < t_1 < t_2 < \dots < t_n = \infty$ the partial sums for the two integrals in question are

$$\sum_{i=1}^{n} E_{t_{i-1}} T[E^{t_i} - E^{t_{i-1}}] = \sum_{i=1}^{n} E_{t_{i-1}} TE^{t_i} - \sum_{i=1}^{n} E_{t_{i-1}} TE^{t_{i-1}}$$

and

$$\sum_{i=1}^{n} [E_{t_{i-1}} - E_{t_i}]TE^{t_i} = \sum_{i=1}^{n} E_{t_{i-1}} TE^{t_i} - \sum_{i=1}^{n} E_{t_i} TE^{t_i}$$

Now the first summation is the same in both cases whereas since $E^{t_0} = 0$ and $E_{t_n} = 0$ the second summations also coincide after a change of the index of summation and dropping the zero terms. Since the partial sums for the two Cauchy integrals coincide so do the integrals and the proof is complete.

Clearly, the same argument used above for uniformly convergent Cauchy integrals also applies to the strongly convergent case.

<u>Lemma</u>: For any bounded operator

$$SRC\int_G dE(t)TE^t = SLC\int_G E_t TdE(t)$$

(i.e., one integral exists if and only if the other exists and their values coincide when they exist.)

We say than an operator, T, is <u>lower triangular</u> if the common value of the integrals

$$RC\int_G dE(t)TE^t = LC\int_G E_t TdE(t)$$

is T, and we say that it is <u>strongly lower triangular</u> if the common value of the integrals

$$SRC\int_G dE(t)TE^t = SLC\int_G E_t TdE(t)$$

is T. (SA-4).

<u>Theorem</u>: The following are equivalent for an operator T.

i) It is causal.

ii) It is lower triangular.

iii) It is strongly lower triangular.

Proof: <u>i) => ii)</u>. If T is causal $TE_t = E_t TE_t$ for all t hence for any partition, $-\infty = t_0 < t_1 < t_2 < \dots < t_n = \infty$ of G we have the partial sum

$$T = TI = T \sum_{i=1}^{n} E^{t_i} - E^{t_{i-1}} = T \sum_{i=1}^{n} E_{\Delta_i} = \overset{n}{\underset{i=1}{\equiv}} T E_{\Delta_i}$$

where

$$E_{\Delta_i} = [E^{t_i} - E^{t_{i-1}}] = E_{t_{i-1}}[E^{t_i} - E^{t_{i-1}}] = E_{t_{i-1}} E_{\Delta_i}$$

Now since T is causal

$$TE_{\Delta_i} = TE_{t_{i-1}}E_{\Delta_i} = E_{t_{i-1}}TE_{t_{i-1}}E_{\Delta_i} = E_{t_{i-1}}TE_{\Delta_i}$$

hence

$$T = \sum_{i=1}^{n} TE_{\Delta_i} = \sum_{i=1}^{n} E_{t_{i-1}}TE_{\Delta_i}$$

Which is precisely in the form of the partial sums for the integral $LC\int_G E_t TdE(t)$ and since these partial sums are all equal to T they converge uniformly to T thereby verifying ii).

ii) => iii). Of course, if an integral converges uniformly it also converges strongly; hence ii) implies iii). Finally, if T is strongly lower triangular then

$$TE_s = SLC\int_G E_t TdE(t)E_s = SLC\int_{[s,\infty)} E_t TdE(t)$$

Now in the range of integration, $t \geq s$, we have $E_t = E_s E_t$; hence

$$TE_s = SLC\int_{[s,\infty)} E_s E_t TdE(t) = E_s[SLC\int_{[s,\infty)} E_t TdE(t)] = E_s TE_s$$

verifying that T is causal and completing the proof.

Clearly, the theorem implies that the concept of strong lower triangularity is redundant since if the integral converges in one sense it also converges in the other.

For the case of a memoryless operator the lower triangularity is immediate for if E_t commutes with T then

$$E_{t_{i-1}}T[E^{t_i}-E^{t_{i-1}}] = T[E^{t_i}-E^{t_{i-1}}]$$

Hence the partial sums for $LC\int_G E_t TdE(t)$ coincide with those for $LC\int_G TdE(t)$

which clearly converges to T. Hence every memoryless operator is, indeed, lower triangular. We note, however, that in this same case

$$E_{t_i}\ T[E^{t_i}-E^{t_{i-1}}] = 0$$

hence

$$RC\int_G E_t TdE(t) = 0$$

for a memoryless operator. The use of the correctly "biased" Cauchy integral is thus basic to the result which could not be obtained using the incorrect Cauchy integral or the Lebesgue integral (which is well defined in the memoryless case though not the general case).

2. Upper Triangularity

A set of results dual to the preceeding apply to anti-causal operators and are stated in the following without proof.

Lemma: For any bounded operator T

$$RC\int_G E^t TdE(t) = LC\int_G dE(t)TE_t$$

Lemma: For any bounded operator T

$$SRC\int_G E^t TdE(t) = SLC\int_G dE(t)TE_t$$

If the common value of the integrals in the first lemma is T we say that T is upper triangular and if the common value of the integrals in the second lemma is T we say that T is strongly upper triangular. As before convergence in either mode implies the other thus rendering the concept of strong upper triangularity redundant. We have:

Theorem: The following are equivalent for an operator T.

i) It is anti-causal.

ii) It is upper triangular.

iii) It is strongly upper triangular.

3. Diagonal Integrals

Results analogous to the preceeding can be obtained for memoryless operators if one replaces the integrals of triangular truncation by diagonal Cauchy integrals (GO-3, BR-1, SA-4). For this purpose we define the (strong) right diagonal Cauchy integral of an operator valued function, f, on G relative to the spectral measure E to be the limit in the (strong) uniform operator topology of the partial sums

$$\Sigma_p = \sum_{i=1}^{n} [E^{t_i}-E^{t_{i-1}}]\, f(t_i)[E^{t_i}-E^{t_{i-1}}]$$

if it exists over the net of all finite partitions of G. In that case the integrals are denoted by

$$RC\int_G dE(t)f(t)dE(t)$$

and

$$SRC\int_G dE(t)f(t)dE(t)$$

in the uniform and strong topologies, respectively. Similarly, by evaluating f at t_{i-1} rather than t_i in our partial sums we obtain the (strong) left diagonal Cauchy integrals

$$LC\int_G dE(t)f(t)dE(t)$$

and

$$SLC\int_G dE(t)f(t)dE(t)$$

in the two topologies.

In many cases, the value of a diagonal Cauchy integral is independent of whether a left or right bias is used; for instance, if f(t) = T is constant. In these cases we denote the common value of these integrals by

$$C\int_G dE(t)f(t)dE(t)$$

and

$$SC\int_G dE(t)f(t)dE(t)$$

for the two modes of convergence.

We say that an operator, T, is <u>diagonal</u> if

$$T = C\int_G dE(t)TdE(t)$$

and we say that it is <u>strongly diagonal</u> if

$$T = SC\int_G dE(t)TdE(t)$$

Analogously to the triangular case we have the following theorem (SA-4).

<u>Theorem</u>: For a bounded operator, T, the following are equivalent.

i) It is memoryless.

ii) It is diagonal.

iii) It is strongly diagonal.

Proof: <u>i) => ii)</u>. If T is memoryless it commutes with E^t; hence

$$[E^{t_i}-E^{t_{i-1}}]T[E^{t_i}-E^{t_{i-1}}] = T[E^{t_i}-E^{t_{i-1}}]$$

whence the partial sums for $C\int_G dE(t)TdE(t)$ coincide with those for $C\int_G TdE(t)$ which converges to T.

ii) => iii). If the diagonal integral converges uniformly it also converges strongly hence ii) implies iii).

iii) => i). If

$$T = SC\int_G dE(t)TdE(t)$$

then

$$TE_s = SC\int_G dE(t)TdE(t)E_s = SC\int_{[s,\infty)} dE(t)TdE(t)$$
$$= SC\int_{[s,\infty)} E_s dE(t)TdE(t) = E_s T$$

since

$$E_s[E^{t_i}-E^{t_{i-1}}] = [E^{t_i}-E^{t_{i-1}}]$$

once the range of integration has been restricted. T is therefore memoryless and the proof is complete.

Consistent with the theorem, the strong convergence of the diagonal integral of T implies its uniform convergence; hence, the concept of a strongly diagonal operator is redundant.

Finally, we note that the three theorems on triangular and diagonal integrals combine to yield a number of alternative representation for the integral of diagonal truncation by combining an upper triangular and lower triangular integral.

E. Strictly Causal Operators

1. Strict Lower Triangularity

Consistent with our characterization of the causal operators as triangular matrices and operators it is natural to attempt to define a class of strictly causal operators which correspond to matrices which are strictly lower triangular (i.e., they are zero both above the diagonal and on the diagonal). Unlike the causal operators which are algebraically defined, unless one makes some restrictive assumption, strict causality is essentially an analytic concept which is not characterizable by any purely algebraic equality. Rather, it is defined wholly in terms of appropriate integrals of strict triangular truncation, this in turn leading to two alternative definitions for strict causality depending on whether the integrals of strict triangular truncation are defined with respect to uniform or strong convergence. Our integrals of strict triangular truncation are characterized by the following lemmas, the proofs of which are similar to those for the corresponding lemmas on the integrals of triangular truncation.

Lemma: For any bounded operator, T,

$$LC\int_G dE(t)TE^t = RC\int_G E_t T dE(t)$$

Lemma: For any bounded operator, T,

$$SLC\int_G dE(t)TE^t = SRC\int_G E_t T dE(t)$$

Consistent with the lemmas we say that a bounded operator T is strictly causal if the common value of the uniformly convergent integrals of the lemma is T and we say that it is strongly strictly causal if the common value of the strongly convergent integrals of the lemma is T. Clearly, strict causality implies strong strict causality.

Although we have no direct algebraic characterization for the strictly causal operators a number of alternative characterizations of the concept are possible via the various integrals of triangular truncation (DE-1).

<u>Theorem</u>: For a bounded operator, T, the following are equivalent.

i) T is strictly causal.

ii) T is causal and

$$RC\int_G E^t T dE(t) = LC\int_G dE(t) T E_t = 0$$

iii) T is causal and

$$C\int_G dE(t) T dE(t) = 0$$

Proof: <u>i) => ii).</u> If T is strictly causal

$$TE_s = RC\int_G E_t T dE(t) E_s = RC\int_{[s,\infty)} E_t T dE(t)$$

$$= RC\int_{[s,\infty)} E_s E_t T dE(t) = E_s[RC\int_{[s,\infty)} E_t T dE(t)] = E_s T E_s$$

showing that T is causal. Now representing the strictly causal T as

$$RC\int_G E_t T dE(t) = T$$

and trivially representing T as

$$RC\int_G T dE(t) = T$$

we have

$$RC\int_G E^t T dE(t) = RC\int_G T dE(t) - RC\int_G E_t T dE(t) = T - T = 0$$

verifying ii)

ii) => iii). Clearly, it suffices to prove that the two conditions of ii) imply that

$$C\int_G dE(t)TdE(t) = 0$$

Now, the partial sums of this integral are of the form

$$\sum_{i=1}^{n} [E^{t_i}-E^{t_{i-1}}]T[E^{t_i}-E^{t_{i-1}}] = \sum_{i=1}^{n} E^{t_i}T[E^{t_i}-E^{t_{i-1}}]$$

$$- \sum_{i=1}^{n} E^{t_{i-1}}T[E^{t_i}-E^{t_{i-1}}]$$

The first of these sums converges to

$$RC\int_G E^t TdE(t) = 0$$

where the equality results from the hypotheses of ii), and the second partial sum converges to

$$LC\int_G E^t TdE(t) = LC\int_G TdE(t)-LC\int_G E_t TdE(t) = T-T = 0$$

hence since both sums converge to zero so does the diagonal integral.

iii) => i). Finally, each partial sum for the integral

$$RC\int_G E_t TdE(t)$$

is of the form

$$\sum_{i=1}^{n} E_{t_i}T[E^{t_i}-E^{t_{i-1}}] = \sum_{i=1}^{n} E_{t_{i-1}}T[E^{t_i}-E^{t_{i-1}}]$$

$$- \sum_{i=1}^{n} [E^{t_i}-E^{t_{i-1}}]T[E^{t_i}-E^{t_{i-1}}]$$

where the two sums converge to

$$LC\int_G E_t T dE(t) = T$$

since iii) implies that T is causal and

$$C\int_G dE(t)TdE(t) = 0$$

via the hypotheses of iii). The partial sums for the integral of strict triangular truncation thus converge to T-0=T and the proof of the theorem is complete.

Clearly, the proof of the preceeding theorem was independent of the mode of convergence employed; hence:

Theorem: For a bounded operator, T, the following are equivalent.

i) T is strongly strictly causal.

ii) T is causal and

$$SRC\int_G E^t T dE(t) = SLC\int_G dE(t)TE_t = 0$$

iii) T is causal and

$$SC\int_G dE(t)TdE(t) = 0$$

Also in the process of proving the theorem we have shown that

$$C\int_G dE(t)TdE(t) = RC\int_G E^t T dE(t) - LC\int_G E^t T dE(t)$$

when the integrals involved are assured to exist. Of course, similar equalities hold with the integrals of triangular truncation replaced by their various equivalents and/or the entire equality replaced with strongly convergent integrals.

Consider the operator, T, on the space $L_2(Z,R,m)$ with its usual resolution of the identity defined by

$$y(k) = \sum_{j=-\infty}^{\infty} \underline{T}(k,j)u(j)$$

Now this is causal if and only if $\underline{T}(k,j) = 0$ for $k < j$ and it is strongly strictly causal if and only if $\underline{T}(k,j) = 0$ for $k \leq j$. To see this we observe that

$$SC\int_G dE(t)TdE(t)$$

is represented by

$$y(k) = \underline{T}(k,k)u(k)$$

where the existence of the strongly convergent integral is assured by the fact that the given representation is well defined for each u in $L_2(Z,R,m)$. Now clearly the diagonal integral is zero if and only if $\underline{T}(k,k) = 0$; hence, our contention is verified. We note that the above argument fails for strictly causal operators since the diagonal integral may fail to converge uniformly (though the representation given for the diagonal integral is valid when it converges uniformly). Hence, unlike the causal operators, the uniformly convergent and strongly convergent strict triangular integrals define distinct classes of operator.

For the space $L_2(R,R,m)$ arguments similar to those required for the above example assure that an operator T represented by

$$y(t) = \int_R \underline{T}(t,q)u(q)dq$$

is strongly strictly causal if $\underline{T}(t,q) = 0$ for $t \leq g$. Similarly, differential and difference operators defined by

$$X(t) = A(t)X(t) + B(t)u(t) \qquad X(0) = 0$$
$$y(t) = C(t)X(t)$$

and

$$X(k+1) = A(k)X(k) + B(k)u(k) \qquad X(0) = 0$$
$$y(k) = C(k)X(k)$$

respectively have solution of the same form as the above strongly strictly causal integral operators and hence are also strongly strictly causal though, in general, they are not strictly causal.

2. Strict Upper Triangularity

Following a development parallel to that used for the strictly causal operators we may define classes of strictly anti-causal operators with dual properties.

Lemma: For any bounded linear operator, T,

$$LC\int_G E^t T dE(t) = RC\int_G dE(t) T E_t$$

Lemma: For any bounded linear operator, T,

$$SLC\int_G E^t T dE(t) = SRC\int_G dE(t) T E_t$$

We say that T is strictly anti-causal if the common value of the above uniformly convergent integrals is T and we say that T is strongly strictly anti-causal if the common value of the above strongly convergent integrals is T. Of course, strict anti-causality implies strong strict anti-causality but as in the strictly causal case the converse is false.

Theorem: For a bounded operator, T, the following are equivalent.

i) T is strictly anti-causal.

ii) T is anti-causal and

$$LC\int_G E_t T dE(t) = RC\int_G dE(t) T E_t = 0$$

iii) T is anti-causal and

$$C\int_G dE(t)TdE(t) = 0$$

Theorem: For a bounded operator, T, the following are equivalent.

i) T is strongly strictly anti-causal.

ii) T is anti-causal and $SLC\int_G E^t EdE(t) = SRC\int_G dE(t)TE_t = 0$

iii) T is anti-causal and $SC\int_G dE(t)TdE(t) = 0$

3. A Lemma

As a preliminary to our closure theorem on strictly causal operators we have the following lemma which will prove to be useful on a number of occasions wherein operator valued integrals relative to spectral measures are employed.

Lemma: Let E be a spectral measure on G, A_i be a disjoint family of Borel Sets and T_i be an arbitrary family of bounded operators. Then

$$||\sum_{i=1}^{n} E(A_i)T_iE(A_i)|| = \sup_i ||E(A_i)T_iE(A_i)||$$

Proof: Since the images of $E(A_i)T_iE(A_i)x$ are orthogonal for all x we have

$$||\sum_{i=1}^{n} E(A_i)T_iE(A_i)x||^2 = \sum_{i=1}^{n} || E(A_i)T_iE(A_i)E(A_i)x||^2$$

$$\leq \sum_{i=1}^{n} ||E(A_i)T_iE(A_i)||^2||E(A_i x||^2$$

$$\leq \sup_i ||E(A_i)T_iE(A_i)||^2 \sum_{i=1}^{n} ||E(A_i)x||^2$$

$$= \sup_i ||E(A_i)T_iE(A_i)||^2||x||^2$$

Thus upon taking the sup over all x of norm one the inequality

$$||\sum_{i=1}^{n} E(A_i)T_iE(A_i)||^2 \leq \sup_i ||E(A_i)TE(A_i)||^2$$

is obtained.

On the other hand if y is of norm less than or equal to one then so is $E(A_j)y = x$ and

$$||\sum_{i=1}^{n} E(A_i)T_iE(A_i)x|| = ||\sum_{i=1}^{n} E(A_iT_iE(A_i)E(A_j)y||$$

$$||E(A_j)T_jE(A_j)y|| = ||E(A_j)T_jE(A_j)x||$$

Now upon taking the sup over all x of norm less than or equal to one it will be greater than or equal to the sup over the subclass of x having the form $E(A_j)y$ hence

$$||\sum_{i=1}^{n} E(A_i)T_iE(A_i)|| \geq ||E(A_j)T_jE(A_j)||$$

Finally, since the inequality holds for all j it must hold when j is chosen to achieve

$$||E(A_j)T_jE(A_j)|| = \sup_i ||E(A_i)T_iE(A_i)||$$

thereby verifying the converse inequality and completing the proof.

4. Operator Algebras

Like the causal operators the strictly causals form a well behaved subset of the space of bounded linear operators, though not as well behaved as the causals (DE-1, SA-4).

Theorem: The space of strictly causal (strictly anti-causal, strongly strictly causal, strongly strictly anti-causal) operators forms a Banach Algebra without identity which is closed in the uniform operator topology of the space of all bounded linear operators. Moreover, the strictly causal (strongly strictly causal) operators form a two sided ideal in the space of causal operators and the strictly anti-causal (strongly strictly anti-causal) operators form a two sided ideal in the space of anti-causal operators.

Proof: The proofs in the four cases are essentially the same; hence, we only consider the strictly causal case. These operators form a linear space since they are defined by a linear equation

$$T = LC\int_G dE(t)TE^t$$

and do not include the identity since the fact that it is memoryless implies that

$$C\int_G dE(t)IdE(t) = I \neq 0$$

To verify that the strictly causals form an ideal in the causals (and hence also that the strictly causals are closed under multiplication since every strictly causal is causal) we let T be causal and S be strictly causal hence ST is causal and it suffices to show that

$$C\int_G dE(t)STdE(t) = 0$$

Now for any partition of G, $-\infty = t_0 < t_1 < t_2 < \dots < t_n = \infty$ we have by the causality of S and T that

$$[E^{t_i}-E^{t_{i-1}}]S = [E^{t_i}-E^{t_{i-1}}]SE^{t_i}$$

and

$$T[E^{t_i}-E^{t_{i-1}}] = E_{t_{i-1}}\, T\, [E^{t_i}-E^{t_{i-1}}]$$

hence

$$[E^{t_i}-E^{t_{i-1}}]ST[E^{t_i}-E^{t_{i-1}}]=[E^{t_i}-E^{t_{i-1}}]SE^{t_i}E_{t_{i-1}}T[E^{t_i}-E^{t_{i-1}}]$$

$$= [E^{t_i}-E^{t_{i-1}}]S[E^{t_i}-E^{t_{i-1}}]T[E^{t_i}-E^{t_{i-1}}]$$

Thus

$$\left|\left| \sum_{i=1}^{n} [E^{t_i}-E^{t_{i-1}}]ST[E^{t_i}-E^{t_{i-1}}]\right|\right|$$

$$=\left|\left|\sum_{i=1}^{n}[E^{t_i}-E^{t_{i-1}}]S[E^{t_i}-E^{t_{i-1}}]T[E^{t_i}-E^{t_{i-1}}]\right|\right|$$

$$= \sup_i ||[E^{t_i}-E^{t_{i-1}}]S[E^{t_i}-E^{t_{i-1}}]T[E^{t_i}-E^{t_{i-1}}]||$$

$$\leq \sup_i ||[E^{t_i}-E^{t_{i-1}}]S[E^{t_i}-E^{t_{i-1}}]||\ ||T||$$

by the lemma. Since S is strictly causal

$$C\int_G dE(t)SdE(t) = 0$$

hence the partial sums of the form

$$\left|\left| \sum_{i=1}^{n} [E^{t_i}-E^{t_{i-1}}]S[E^{t_i}-E^{t_{i-1}}]\right|\right| = \sup_i ||[E^{t_i}-E^{t_{i-1}}]S[E^{t_i}-E^{t_{i-1}}]||$$

converge to zero hence so do the partial sums for

$$C\int_G dE(t)STdE(t)$$

thereby verifying that ST is strictly causal.

Finally, we must verify the closure condition. If the T_i are causal and converge uniformly to T then T is causal and we must show that

$$C\int_G dE(t)TdE(t) = 0$$

For any $\varepsilon > 0$ choose j such that

$$||T_j - T|| < \varepsilon/2$$

and a partition of the group G $-\infty = t_0 < t_1 < t_2 < \ldots < t_n = \infty$ such that

$$||\sum_{i=1}^{n} [E^{t_i}-E^{t_{i-1}}]T_j[E^{t_i}-E^{t_{i-1}}]|| < \varepsilon/2$$

Here it is always possible to find such a j since T_i converges uniformly to T and it is possible to find the partition since the fact that T_j is strictly causal implies that

$$C\int_G dE(t)T_j dE(t) = 0$$

Now, we have

$$||\sum_{i=1}^{n} [E^{t_i}-E^{t_{i-1}}]T[E^{t_i}-E^{t_{i-1}}]||$$

$$=||\sum_{i=1}^{n} [E^{t_i}-E^{t_{i-1}}](T-T_j)[E^{t_i}-E^{t_{i-1}}]+\sum_{i=1}^{n} [E^{t_i}-E^{t_{i-1}}]T_j[E^{t_i}-E^{t_{i-1}}]||$$

$$\leq \sup_i ||[E^{t_i}-E^{t_{i-1}}](T-T_j)[E^{t_i}-E^{t_{i-1}}]||+\sup_i ||\sum_{i=1}^{n} [E^{t_i}-E^{t_{i-1}}]T_j[E^{t_i}-E^{t_{i-1}}]||$$

$$\leq \varepsilon/2+\varepsilon/2 = \varepsilon$$

showing that the partial sums for

$$C\int_G dE(t)TdE(t)$$

converge to zero and that T is strictly causal.

Note that unlike the case for causal operators the strictly causal operators are not closed in the strong operator topology. For instance, the shift operators of $L_2(R,R,m)$ are strongly strictly causal for $t > 0$ but converge strongly to the identity. Also note that the theorem implies that the strictly causal operators are an ideal of the strongly strictly causal operators since both are ideals of the causal operators and one is contained in the other.

5. Adjoints

The relationship between the strictly causal and strictly anti-causal operators is given by the following theorem.

Theorem: An operator T is (strongly) strictly causal if and only if T* is (strongly) strictly anti-causal.

Proof: Since the partial sums for

$$LC\int_G E^t T^* dE(t)$$

and

$$LC\int_G dE(t) T E^t$$

are adjoints of one and another and the adjoint is a linear isometry on the space of bounded linear operators on a Hilbert space the first integral converges to T* if and only if the second converges to T thereby verifying the contention of the theorem.

F. Operator Decomposition

1. Additive Decomposition

Since we have defined a number of classes of special types of operator it is natural to attempt to represent an arbitrary operator as a sum or product of operators from these classes, these decompositions proving to be of fundamental importance in many system theoretic applications (PR-3, PR-4, SA-4, SA-6, SA-9). The most natural type of decomposition theorem might simply be to write an arbitrary operator as the sum of a causal and and anti-causal operator. Although this can often be done it will not be unique since a memoryless term could be placed in either the causal or anti-causal part. We therefore attempt to decompose an operator as the sum of a strictly causal, strictly anti-causal and memoryless operator, the decomposition being unique when it exists (SA-4, DE-1).

Theorem: Let T be an arbitrary linear bounded operator. Then T can be decomposed as

$$T = C + M + A$$

where C is strictly causal, M is memoryless and A is strictly anti-causal if and only if two of the three integrals

$$RC\int_G E_t T dE(t)$$

$$LC\int_G E^t T dE(t)$$

and

$$C\int_G dE(t) T dE(t)$$

exist in which case the third also exists and the decomposition is given by

$$C = RC\int_G E_t T dE(t)$$

$$A = LC\int_G E^t T dE(t)$$

and

$$M = C\int_G dE(t) T dE(t)$$

Moreover, when the decomposition exists it is unique.

Proof: First assume that all three integrals exist. Then

$$T = RC\int_G T dE(t) = RC\int_G [E_t + E^t] T dE(t)$$

$$= RC\int_G E_t T dE(t) + RC\int_G E^t T dE(t)$$

$$= RC\int_G E_t T dE(t) + LC\int_G E^t T dE(t) + C\int_G dE(t) T dE(t)$$

$$= C + M + A$$

where all of the integrals exist by hypothesis and in fact since $RC\int_G TdE(t)$ trivially exists the equality itself assures that if any two of the integrals will exist then so does the third. Now

$$RC\int_G E_t C dE(t) = RC\int_G E_t [RC\int_G E_s T dE(s)] dE(t)$$

$$= RC\int_G E_r T dE(r) = C$$

via the idempotent property of the integral hence C is strictly causal and by a similar argument A is strictly anti-causal and M is memoryless. We thus have the required decomposition.

Conversely, if the decomposition exists since C is strictly causal

$$C = RC\int_G E_t C dE(t)$$

while since A and M are anti-causal; being strictly anti-causal and memoryless, respectively;

$$RC\int_G E_t(A+M)dE(t) = RC\int_G (I-E^t)(A+M)dE(t)$$

$$= RC\int_G (A+M)dE(t) - RC\int_G E^t(A+M)dE(t) = (A+M)-(A+M) = 0$$

We thus have

$$RC\int_G E_t TdE(t) = RC\int_G E_t(C+M+A)dE(t)$$

$$= RC\int_G E_t CdE(t) + RC\int_G E_t(M+A)dE(t) = C + 0 = C$$

showing that C is indeed of the required form. Of course, a similar argument will reveal that A and M are of the required form. Note that all of the above integrals are assured to exist by the various triangular characterizations we have thus far obtained for causal and anti-causal operators.

Finally, we must verify the uniqueness of the decomposition. Assume we are given two decompositions

$$T = C + M + A = \underline{C} + \underline{M} + \underline{A}$$

thus

$$C - \underline{C} = \underline{M} + \underline{A} - M - A$$

is strictly causal (since C - $\underline{C}$ is strictly causal) and also anti-causal (since M, A, $\underline{M}$ and $\underline{A}$ are all anti-causal). We thus have

$$0 = RC\int_G E^t(C-\underline{C})dE(t) = RC\int_G E^t(\underline{M}+\underline{A}-M-A)dE(t)$$

$$= \underline{M} + \underline{A} - M - A = C - \underline{C}$$

showing that $C = \underline{C}$ and verifying the uniqueness of the strictly causal term in the decomposition. Similarly, the uniqueness of the other terms may be verified which completes the proof of the theorem.

2. Strong Additive Decomposition

Clearly, the results of the preceeding theorem are independent of the mode of convergence employed; hence:

Theorem: Let T be an arbitrary linear bounded operator. Then T can be decomposed as

$$T = C + M + A$$

where C is strongly strictly causal, M is memoryless and A is strongly strictly anti-causal if and only if two of the three integrals

$$SRC\int_G E_t T dE(t)$$

$$SLC\int_G E^t T dE(t)$$

and

$$SC\int_G dE(t) T dE(t)$$

exist in which case the third also exists and the decomposition is given by

$$C = SRC\int_G E_t T dE(t)$$

$$A = SLC\int_G E^t T dE(t)$$

and

$$M = SC\int_G dE(t) T dE(t)$$

Moreover, when the decomposition exists it is unique.

Consider an operator on $l_2 = L_2(Z,R,m)$, with the usual resolution of the identity, defined by the convolution

$$y(k) = \sum_{j=-\infty}^{\infty} \underline{T}(k-j)u(j)$$

Now the operators C, M and A defined by

$$y(k) = \sum_{j=-\infty}^{k-1} \underline{T}(k-j)u(j)$$

$$y(k) = \underline{T}(0)u(k)$$

and

$$y(k) = \sum_{j=k+1}^{\infty} \underline{T}(k-j)u(j)$$

respectively, add to T and are strongly strictly causal, memoryless and strongly strictly anti-causal, respectively. As such that constitutes the required strongly convergent decomposition. In general, however, C may not be strictly causal in which case the required decomposition will fail to exist in the uniform sense. For instance, if

$$\underline{T}(i) = 1/i$$

C and A will be unbounded and hence cannot be the uniform limit of the partial sums for the integrals of strict triangular truncation (LA-1). In fact, it is possible to exhibit bounded (and even compact) operators for which the required integrals of strict triangular truncation fail to exist even in the strong topology hence even "nice" operators may fail to have an additive decomposition (DE-1, GO-3). Finally, one special class of convolution operators wherein the above C, M and A are assured to exist in the uniform sense is the case where $\underline{T}$ is an L_1 sequence, hence the required decomposition does exist given reasonable assumptions on the operator T (SA-7).

The required integrals of triangular truncation have been shown to exist for a number of classes on compact operators in the study of the spectral theory of these operators. The largest such class is S_Ω as defined in reference GO-1. A subclass of S_Ω for which the verification of the existence of the integrals of triangular truncation is straightforward is the Hilbert-Schmidt operators wherein the integrals of strict triangular truncation reduce to the orthogonal projections (since these operators form a Hilbert space) onto the strictly causal and strictly anti-causal operators and hence are assured to be well-defined for all Hilbert-Schmidt operators (DE-1). The general proof that the integrals of strict triangular truncation exist for all of S_Ω is, however, quite involved (GO-3).

3. Multiplicative Decomposition

Rather than decomposing an operator as a sum of operators with desirable causality properties we may decompose it as a product of operators with desirable causality properties. To this end we have the following lemma which plays a fundamental role in a number of system theoretic results as well as the derivation of the multiplicative decomposition theorem (SA-3, SA-4):

Lemma: Let T be a contraction on a resolution space (H,E). Then there exists a resolution space $(\underline{H},\underline{E})$ and an isometric extension Σ of T defined on $(H,E)\oplus(\underline{H},\underline{E})$ with matrix representation

$$\Sigma = \left[\begin{array}{c|c} T & 0 \\ \hline \Sigma_{21} & \Sigma_{22} \end{array}\right]$$

where Σ_{21} is a causal operator mapping (H,E) to ($\underline{H}$,$\underline{E}$) and Σ_{22} is a causal operator mapping ($\underline{H}$,$\underline{E}$) to ($\underline{H}$,$\underline{E}$).

Proof: We use precisely the same extension space and operator as for the extension of T^1 of a contractive semi-group(C.C.3). That is, we let D be the range of $(I-T^*T)^{1/2}$ and let $\underline{H}$ be countably many copies of D. Now for any element of $H\oplus\underline{H}$, $(h, d_1, d_2, \ldots)$,we define

$$\Sigma(h,d_1,d_2,d_3,\ldots) = (Th, (I-T^*T)^{1/2}h,d_1,d_2,d_3,\ldots)$$

$$||\Sigma(h,d_1,d_2,\ldots)||^2=||(Th,(I-T^*T)^{1/2}h,d_1,d_2,d_3,\ldots)||^2$$

$$=||Th||^2+||(I-T^*T)^{1/2}h||^2+||\sum_{i=1}^{\infty} d_i||^2$$

$$=<Th,Th>+<(I-T^*T)^{1/2}h,(I-T^*T)^{1/2}h>+\sum_{i=1}^{\infty}||d_i||^2$$

$$=<Th,Th>+<h,(I-T^*T)h>+\sum_{i=1}^{\infty}||d_i||^2$$

$$=<Th,Th>+<h,h>-<Th,Th>+\sum_{i=1}^{\infty}||d_i||^2$$

$$=||h||^2+\sum_{i=1}^{\infty}||d_i||^2=||(h,d_1,d_2,d_3,\ldots)||^2$$

verifying that the operator is indeed isometric. Finally, we must define a spectral measure on $\underline{H}$ which renders Σ_{21} and Σ_{22} causal. To this end let us define D_t to be the image of H_t under $(I-T^*T)^{1/2}$ in D and $\underline{H}_t$ to be countably many copies of D_t in $\underline{H}$. Now we define the resolution of the identity, $\underline{E}^t$ by $\underline{E}^t=I-\underline{E}_t$ where $\underline{E}_t$ is the projection onto $\underline{H}_t$. The fact that this is a well defined resolution of the identity follows from the fact that E^t is a resolution of the identity and that $(I-T^*T)^{1/2}$ maps H onto D. Finally, we let $\underline{E}$ be the spectral measure defined by $\underline{E}^t$. Now for any h in H Σ_{21} has the form

$$\Sigma_{21}h = ((I-T^*T)^{1/2}h,0,0,\ldots)$$

hence if h is in H_t $(I-T^*T)^{1/2}h$ is in D_t and clearly 0 is in D_t, thus $\Sigma_{21}h$ is in $\underline{H}_t$ verifying that Σ_{21} is causal. Similarly, for $\underline{h}=(d_1,d_2,d_3,\ldots)$

$$\Sigma_{22}\underline{h} = (0,d_1,d_2,d_3,\ldots)$$

whence if $\underline{h}$ is in $\underline{H}_t$ each d_i is in D_t and $\Sigma_{22}\underline{h}$, being the orthogonal sum of such d_i and 0, is also in $\underline{H}_t$ verifying that Σ_{22} is causal. Finally, for this map Σ_{12} is trivially zero thus the proof is complete.

Finally, our desired multiplicative decomposition theorem follows from the lemma.

<u>Theorem</u>: Let T be a bounded positive hermitian operator on a resoluiton space (H,E). Then there exists a resolution space $(\underline{H},\underline{E})$ and a causal operator, C, mapping (H,E) into $(\underline{H},\underline{E})$ such that

$$T = C^*C$$

Proof: Clearly it suffices to consider the case where T is a contraction since the result for the general case can be obtained from the contractive result by scaling. Now let

$$S = (I-T)^{1/2}$$

which is a well defined positive hermitian operator when T is a positive hermitian contraction. Let

$$\Sigma = \left[\begin{array}{c|c} (I-T)^{1/2} & 0 \\ \hline \Sigma_{21} & \Sigma_{22} \end{array}\right]$$

be the isometric extension of the lemma for which we have

$$I = \Sigma^*\Sigma = \left[\begin{array}{c|c} (I-T)^{1/2} & \Sigma_{21}^* \\ \hline 0 & \Sigma_{22}^* \end{array}\right] \left[\begin{array}{c|c} (I-T)^{1/2} & 0 \\ \hline \Sigma_{21} & \Sigma_{22} \end{array}\right] = \left[\begin{array}{c|c} I & 0 \\ \hline 0 & I \end{array}\right]$$

Now the upper right hand equality above is

$$I=(I-T)^{1/2}(I-T)^{1/2}+\Sigma_{21}^*\Sigma_{21}=I-T+\Sigma_{21}^*\Sigma_{21}$$

which upon letting $C = \Sigma_{21}$ and rearranging becomes

$$T = C^*C$$

as required to complete the proof of the theorem.

Note that any operator which can be decomposed as $T=C^*C$ must be positive and hermitian; hence, the theorem is really necessary and sufficient. Unlike the additive decomposition one cannot readily convert the above existence theorem into an explicit formula for computing C (some representation of C) given T (some representation of T). Typically, one must solve an equation in terms of a representation of T in order to construct a representation of C. For instance, in the context of various representations one solves a Wiener-Hopf equation (SO-1), factors a complex valued function into functions with specified analyticity characteristics (MS-5, WE-1), or solves a matrix Riccati equation (AN-1).

G. Problems and Discussion

1. Causal Operators

Of the various commonly encountered operators the one whose causality character is least obvious is the derivative (defined on a dense subset of $L_2 = L_2(R,R,m)$). Depending on whether one uses a right, left or two sided definition, $\dot{f}(t_o)$ may depend on values of $f(t)$ after t_o, before t_o of both before and after t_o. In any event, however, it depends only on the values of $f(t)$ in an arbitrarily small neighborhood of t_o. In fact,

> Problem: Show that the derivative is a memoryless operator on $L_2(R,R,m)$ with its usual resolution of the identity.

In essence, the memory of a derivative is arbitrarily short (though non-zero) and is thus memoryless by our definition.

Though the derivative is memoryless it is unbounded and thus the closure theorem for memoryless bounded operators is not applicable. In fact, it is possible to construct non-memoryless operators as the limit of a sequence of unbounded operators.

> Problem: Give an example of a non-memoryless operator on $L_2(R,R,m)$ which is the limit of a sequence of memoryless unbounded operators.

Although most results for causal operators hold in general if one goes to a sufficiently specialized space, additional results can often be obtained. One such space is $l_2^+ = L_2(Z,R,m^+)$ where $m^+(A)$ is equal to the number of positive integers in A (LA-1). This space is characterized by the fact that H^t, t in Z, is finite dimensional (though the whole space is infinite dimensional).

> Problem: Show that every causal operator on $L_2(Z,R,m^+)$ with its

usual resolution structure is bounded.

Another result applicable to a specialized space is the corollary to the Paley-Wiener theorem (KA-1) which states that no ideal filter is causal.

<u>Problem</u>: Show that the operator on $L_2(R,R,m)$ with its usual resolution structure defined by

$$(F_A f)(t) = \frac{1}{2\pi} \int_A [\int_R f(q) e^{-i\omega q} dq] e^{i\omega t} d\omega$$

for any non-null Borel set, $A \neq R$, is not causal.

The concept of strict causality is, physically, somewhat ambiguous and therefore can be given a number of distinct mathematical interpretations as we have already seen via our study of the relationships between strong strict causality and strict causality. Clearly, we could define <u>weak strict causality</u> by taking the limit for the integrals of triangular truncation in the weak operator topology which would yield a class of operators larger than the strongly strictly causal operator (since strong convergence implies weak convergence).

<u>Problem</u>: Determine whether or not the strongly strictly causal operators are properly contained in the weakly strictly causal operators.

In fact, it is possible to define a strict causality concept independently of the integral of strict triangular truncation (WL-3) by requiring that T be causal and that for all t in G for any $\varepsilon > 0$ and $s \leq t$, there exists $r > 0$ in G such that for any x, y in H such that $E^s x = E^s y$ then

$$||E^{s+r}(Tx-Ty)|| \leq \varepsilon ||E^{s+r}(x-y)||$$

<u>Problem</u>: Determine the relationship, if any, between the above strict causality concept and the three strict causality concepts

defined via the integral of strict triangular truncation.

Although a rather large body of literature exists on causal operators the theory is by no means complete. One open problem is the following.

Problem: Under what conditions is a strictly causal operator assured to be compact?

Intuitively, a strictly causal operator is one which has a "transmission" delay though as we have defined the concept the delay may be arbitrarily short so long as the norm of that part of the output with small delay is small. An alternative approach to the concept is to require that there is actually zero transmission for a period of time after the input is applied (WL-3). Such operators are said to have strict delay and are formally defined by the equality

$$E^t T = E^t T E^{t-d}$$

where d is a time delay greater than zero. Of course, a similar definition applies to strictly predictive operators. Although, upon first impression, one would expect that any operator with strict delay is strictly causal, the integrals required for strict causality sometime fail to converge and hence the two concepts are not as closely related as one might expect.

Problem: Show that every operator with strict delay is causal.

Problem: Show that every operator with strict delay is strongly strictly causal if G is an archimedean ordered group.

Problem: Give a set of necessary and sufficient conditions for an operator with strict delay to be strictly causal (strongly strictly causal) when G is an arbitrary ordered group.

For a resolution space (H,E) a gap is a pair (t_1,t_2) of elements of G such that $E^{t_1} \neq E^{t_2}$ and for any t such that $t_1 \leq t \leq t_2$ either

$E^t = E^{t_1}$ or $E^t = E^{t_2}$ (GO-3). The space $L_2(R,K,m)$ with its usual resolution of the identity has no gaps whereas for $L_2(Z,K,m)$ with its usual resolution structure (t_i,t_{i+1}) is a gap for each i.

Problem: Show that in a resolution space with no gaps

$$C\int_G dE(t)TdE(t) = 0$$

for every compact operator T.

Problem: Show that in a resolution space with no gaps the only compact memoryless operator is 0.

Problem: Show that in a resolution space with no gaps that a compact operator is causal (anti-causal) if and only if it is strictly causal (strictly anti-causal).

2. Causal Nonlinear Operators

Unlike the linear case, when one is dealing with nonlinear operators the various equivalent conditions for causality and strict causality diverge and we have several modes of causality. We therefore term the usual definition of causality same-input same-output causality, the condition that $||E^tTx|| \leq ||TE^tx||$ for all t in G and x in H norm causality, and the condition that $E^tTx = 0$ if $E^tx = 0$ zero-input zero-output causality (WI-1).

Problem: Give examples to show that the three causality modes are distinct for nonlinear operators.

Problem: Show that the following are equivalent for a possibly nonlinear operator T.

i) T is same-input same-output causal

ii) $E^tT = E^tTE^t$ for all t in G.

Problem: Show that the following are equivalent for a possibly nonlinear operator T.

i) T is zero-input zero-output causal

ii) H_t is an invariant subspace of T for all t in G.

iii) $TE_t = E_t TE_t$

iv) $E^t TE_t = 0$ for all t in G.

Although the three causality concepts are, indeed, distinct in the nonlinear case they are related (WI-1).

Problem: Show that same-input same-output causality implies norm causality.

Problem: Show that norm causality implies zero-input zero-output causality if T(0) = 0.

Similarly, to the causal case the various equivalent conditions for strict causality are distinct in the nonlinear case (DE-1). We continue to use the term strict causality for the case when

$$T = RC\int_G E_t TdE(t)$$

and refer to an operator which is causal and satisfies

$$C\int_G dE(t)TdE(t) = 0$$

as strongly causal. (Clearly, we could also define strongly strongly causal operators by using a strongly convergent integral. The ambiguity in notation results from the fact that strict and strong causality were originally defined in a uniformly convergent sense, the importance of the strongly convergent case not being realized until the notation was established (DE-1, SA-5)).

Problem: Give examples to show that strong causality and strict causality are distinct for nonlinear operators.

Problem: Determine the relationship, if any, between the strongly causal and strictly causal operators.

Problem: Show that the strongly causal, bounded operators form an ideal in the same-input same-output causal, Lipschitz continuous and bounded operators.

Finally, one additional class of time-dependent operators occurs in the nonlinear case which has no analog in the linear case (DE-1, DE-6). These are termed crosscausal if $E^t x = 0$ implies that $E^t Tx = 0$, $E_t x = 0$ implies that $E_t Tx = 0$ and

$$C\int_G dE(t)TdE(t) = 0$$

and strongly crosscausal if

$$0 = RC\int_G dE(t)TE^t = LC\int_G dE(t)TE_t = C\int_G dE(t)TdE(t)$$

Problem: Show that every strongly crosscausal operator is crosscausal.

The fact that crosscausality is degenerate in the linear case may be verified from the result of the following problem.

Problem: A linear operator is crosscausal if and only it if is the zero operator.

The nonlinear crosscausal operators are, however, well behaved as per the following problem.

Problem: Show that the space of bounded crosscausal operators forms a Banach algebra which is closed in the uniform operator topology of the space of all bounded operators.

In the nonlinear case our additive decomposition theorem has four, rather than three, terms; a memoryless operator, a strongly causal operator, a strongly anti-causal operator (defined in the obvious manner) and a strongly crosscausal operator (DE-1, DE-6).

<u>Problem</u>: Show that if an operator T can be decomposed as

$$T = C + M + A + X$$

where C is strongly causal, A is strongly anti-causal, M is memoryless and X is strongly crosscausal then the decomposition is unique.

<u>Problem</u>: Give an example of a strongly crosscausal operator on $L_2(R,R,m)$.

<u>Problem</u>: Give an example of a crosscausal operator on $L_2(R,R,m)$ which is not strongly crosscausal.

In a number of contexts in the study of nonlinear operators one desires to determine the causality properties of the inverse of an operator (DA-1, WL-1, WL-3). For the cases of zero-input zero-output and same-input same-output causality we have:

<u>Problem</u>: Let T be an arbitrary invertible nonlinear operator. Show that T^{-1} is zero-input zero-output causal if and only if $E^t x \neq 0$ implies $E^t Tx \neq 0$.

<u>Problem</u>: Let T be an arbitrary invertible nonlinear operator. Show that T^{-1} is same-input same-output causal if and only if $E^t x \neq E^t y$ implies that $E^t Tx \neq E^t Ty$.

<u>Problem</u>: Give a set of necessary and sufficient conditions on an invertible nonlinear operator, T, to assure that T^{-1} will be norm causal.

As far as sufficient conditions for causal invertibility are concerned we say that a nonlinear operator is *(Lipschitz) definite* if for every x and y, $x \neq y$

$$\langle x-y, T(x-y) \rangle \neq 0$$

Conjecture: The inverse of every same-input same-output causal (Lipschitz) definite nonlinear operator is same-input same-output causal.

3. Weakly Additive Operators

Although the different forms of causality are distinct for general nonlinear operators there is one class of nonlinear operator for which the linear results hold (DE-1, DE-6, GE-1, WI-1). These are termed *weakly additive* operators and have essentially linear time behavior. They are defined by the requirement that $Tx = TE^t x + TE_t x$ for all t and x.

Problem: Show that an operator of the form $T = LM$ where L is linear and M is a possible nonlinear memoryless operator is weakly additive.

Note that the above result implies that all linear operators and all memoryless nonlinear operators are weakly additive.

Problem: Show that for any weakly additive operator, W, and memoryless nonlinear operator M that $T = WM$ is also weakly additive.

Problem: Show that for any weakly additive operator, W, and linear operator, L, that $T = LW$ is also weakly additive.

Problem: Show that for every weakly additive operator, W, $W(0) = 0$.

The main reason for dealing with weakly additive operators is that the various causality concepts are all equivalent for this class of operators (WI-1).

Problem: For a weakly additive operator, W, show that the following are equivalent.

i) W is same-input same-output causal.

ii) W is norm causal.

iii) W is zero-input-zero-output causal.

Problem: For a weakly additive operator show that it is strictly causal if and only if it is strongly causal.

As in the linear case the crosscausal operators are degenerate in the weakly additive context.

Problem: Show that a weakly additive operator is crosscausal if and only if it is the zero operator.

Consistent with the above, the additive decomposition for a weakly additive operator, like a linear operator, has only three terms.

One interesting, though as yet unsolved, problem associated with the weakly additive operators is their representation in terms of more classical operators. We have the following conjecture.

Conjecture: Let W be a weakly additive operator on a resolution space, (H,E). Then there exists a resolution space $(\underline{H},\underline{E})$, a memoryless (possibly nonlinear) operator, M, mapping (H,E) into $(\underline{H},\underline{E})$ and a linear operator, L, mapping $(\underline{H},\underline{E})$ into (H,E) such that W = LM.

4. J-Spaces

An alternative approach to the causality concepts studied here in resolution space is to work with J-spaces (LE-1, LE-4), that is a triple (H,J,G) (usually written as (H,J)) where H is a Hilbert space, G is an (LCA) topological group and J is a strongly continuous operator (on H) valued function defined on G satisfying

i) $J(t)^* = J(t)$ for all t in G

ii) $J(t)^2 = I$

iii)

$$\underset{t \to \infty}{\text{s-limit}}\, J(t) = -I$$

and

$$\underset{t \to \infty}{\text{s-limit}}\, J(t) = I$$

Although formally different there is a one-to-one correspondence between J spaces and resolution spaces which allows one to do anything in one space that can be done in the other.

Problem: Show that if (H,E) is a resolution space then (H,J) where

$$J(t) = [E^t - E_t]$$

is a J-space.

Problem: Show that if (H,J) is a J-space then (H,E) where

$$E^t = \frac{1}{2}[I + J(t)]$$

is a resolution space.

The transformations defined in the above problems define an equivalence between the categories of resolution spaces and J space which we term the natural transformation. This may then be used to translate our various resolution space concepts to J-space. For instance, we say that an operator T on a J-space (H,J) is causal if

$$J(t)T - TJ(t) = J(t)TJ(t) - T;\ t \text{ in } G$$

which is, in a sense, a condition on the commutant of T and J(t).

Problem: Let (H,E) and (H,J) be naturally equivalent resolution and

J-spaces and let T be an operator on H. Then show that T is causal on (H,E) if and only if T is causal on (H,J).

Of course, all of our other resolution space concepts can be translated to J-spaces via similar formulations. Hopefully, some are simpler in one formulation and some in the other. In fact, rather than defining our Cauchy integrals with respect to E^t we may define them via J(t) and thus our Cauchy integral techniques also carry over to J-spaces.

<u>Problem</u>: Let $C\int$ denote any of the Cauchy integrals and show that for any operator valued function T and interval I in G that

$$C\int_I T(t)dJ(t) = C\int_I 2T(t)dE(t)$$

where (H,E) is a resolution space and (H,J) is its natural transformation.

2. FEEDBACK SYSTEMS

Possibly the most fully developed system theoretic application of resolution space concepts is to the stability problem for feedback systems. Although most of the results were originally formulated in a function space setting, the techniques employed are resolution space oriented, and hence readily translated into the resolution space notation in which they are presented in the present chapter. Early workers in the field include Willems (WL-1, WL-3), Zames (ZA-1, ZA-2, ZA-3), Sanberg (SN-2), Damborg and Naylor (DA-1, DA-2, DA-3) and DeSantis (DE-1, DE-3, DE-5). Closely related to the feedback system stability problem is its sensitivity problem for which we present the formulation of Porter and Zahm (PR-4) also translated into the resolution space notation of the present monograph.

We begin with a study of the concept of well-posedness (WL-3) which essentially amounts to the determination of whether or not a feedback system has a unique solution for each input. In general, this is not the case. If, however, one works in an appropriate extension space of the given resolution space such solutions may be shown to exist for many practical classes of feedback system. We then consider the problem of stability analysis for well-posed systems showing that this reduces to the requirement that the feedback system have (in an appropriate sense) unique solutions in the given resolution space for each input in the given resolution space. As such, even though one must go to an extension space to study well-posed feedback systems the study of feedback systems which are simultaneously well-posed and stable may be carried out in the given resolution space. The extension space (ZA-1, ZA-2, SN-1, WL-1, WL-3) and normed space (DA-1, DA-2, DA-3, DE-1, DE-3, DE-5) approaches to the stability problem are thus equivalent and the various normed space sufficient condition for stability may be formulated for our extension space approach to stability. Finally, we consider the problem of feedback system sensitivity developing a criterion for the comparison of equivalent open loop and feedback system; the theory being illustrated for the particular case of equivalent open loop and feedback controllers.

A. Well-Posedness

1. Feedback Equations

A feedback system is characterized by the block diagram shown below or alternatively the set of equations

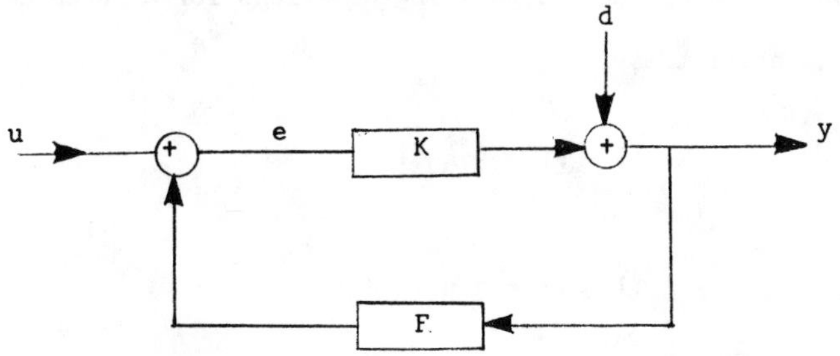

Figure: Block Diagram for Feedback System.

$$y = Ke + d$$

and

$$e = Fy + u$$

where K and F are causal operators on a fixed resolution space (H,E) (over a group G) which are, respectively, termed the plant and feedback operators. The concept has a number of variations with additional operators or "signals" included which will be invoked on occasion. The most interesting properties of a feedback system, however, manifest themselves in the configuration shown above and hence we will restrict ourselves to this case where possible.

Conceptually, we view u as the system input "signal," d as a disturbance in the output of the system (due possibly to initial conditions or noise); these two signals being interpreted as the independent variables of the feedback system. Similarly, the system dependent variables are e and y where y is interpreted as the system output and e as the input to the plant K.

Upon combining the two defining equations for a feedback system we obtain the equalities

$$(I - KF)y = Ku + d$$

and

$$(I - FK)e = u + Fd$$

relating the two dependent variables to the two independent variables. Clearly, for y and e to be uniquely determined by u and d we require that the inverses of (I - KF) and (I - FK) exist. Here (I - KF) is termed the <u>return difference</u> and KF the <u>open loop gain</u> of the feedback system though in fact we could equally well have applied these terms to (I - FK) and FK since their properties are essentially identical (SA-10, SI-1).

<u>Theorem</u>: For causal operator, K and F, on a resolution space (H,E) $(I - KF)^{-1}$ exists if and only if $(I - FK)^{-1}$ exists. Moreover, one of these inverses is causal if and only if the other is causal,

$$(I - KF)^{-1}K = K(I - FK)^{-1}$$

and

$$(I - FK)^{-1}F = F(I - KF)^{-1}$$

The theorem results from the standard (if not well known) algebraic equality (SA-10, SI-1)

$$(I - XY)^{-1} = I - X(I + YX)^{-1}Y$$

with the appropriate substitutions for X and Y and the details will not be given.

Consistent with the theorem our operators relating y and e to u and d are given by

$$\begin{aligned} y &= (I - KF)^{-1}Ku + (I - KF)^{-1}d \\ &= K(I - FK)^{-1}u + (I - KF)^{-1}d \end{aligned}$$

and

$$\begin{aligned} e &= (I - FK)^{-1}u + (I - FK)^{-1}Fd \\ &= (I - FK)^{-1}u + F(I - KF)^{-1}d \end{aligned}$$

when the inverses exist. In general, however, there is no reason to expect that these inverses exist. Surprisingly, however, these inverses are well defined for most commonly encountered feedback system if one deals with an appropriate algebraic extension of H rather than H itself. Of course, the above theorem, being algebraic in nature, is applicable in such cases.

2. Extension Spaces

An algebraic spectral measure is the natural generalization of a spectral measure to a vector space with those defining properties for the spectral measure which are Hilbert space in nature being deleted. For this purpose if $\hat{H}$ is a (not necessarily normed) vector space, a set function, $\hat{E}$, defined on the Borel sets of G whose values are linear operators on $\hat{H}$ satisfying

i) $\hat{E}(G) = I$.

ii) $\hat{E}(\emptyset) = 0$.

iii) $\hat{E}(A)\hat{E}(B) = \hat{E}(A \cap B)$ for all Borel sets A and B.

iv) $\hat{E}$ is finitely additive.

is termed an <u>algebraic spectral measure</u>. Following the standard notation for spectral measures we denote $\hat{E}(-\infty,t)$ by $\hat{E}^t$ and $\hat{E}(t,\infty)$ by $\hat{E}_t$. As in the case of a spectral measure on a Hilbert space the classical example of an algebraic spectral measure on a vector space of functions or sequences is the operation of multiplication by the characteristic function of a set, this being termed the <u>usual algebraic spectral measure</u> for a function space.

For a resolution space (H,E) an <u>extension space</u> $(\hat{H},\hat{E})$ is a vector space $\hat{H}$ together with an algebraic spectral measure, $\hat{E}$, on $\hat{H}$, such that:

i) $H \subset \hat{H}$.

ii) $\hat{E}$ restricted to H is equal to E.

iii) The range of $\hat{E}^t$ is in H for all t in G.

iv) Given any family of vectors in H, x^c, c in C; where C is any co-final subset of G (i.e., for any t in G there exists c in C such that $c \geq t$ in the ordering of G) such that

$$E^s x^t = x^s$$

whenever $s \leq t$, s and t in C. Then there exists a unique vector $\hat{x}$ in $\hat{H}$ such that

$$x^c = \hat{E}^c \hat{x}$$

for all c in C.

Note that the last condition here is of primary importance since it implies that the properties of the vectors in $\hat{H}$ are entirely determined by their truncates in H.

For the space $L_2(G,K,\mu)$ where K is a Hilbert space with its usual spectral measure, the space of K valued functions defined on G such that

$$\int_{(-\infty,t)} ||f(q)||^2 d\xi(q) < \infty$$

for all t form an extension space with $\hat{E}(A)$ taken to be multiplication (in $\hat{H}$) by the characteristic function of A. In fact, this space is the usual model for the extension space concept (WL-3). Similarly, the set of all locally integrable functions on G which are zero for t less than some t_o form an extension space for the square integrable functions which are zero for t less than t_o. In particular, the set of all sequences on the positive integers extend $L_2(Z,K,m^+)$ (LA-1, DS-1). Of course (H,E) is an extension of itself for any resolution space.

Although the extension space $\hat{H}$ is not explicitly normed we can partially circumvent the difficulties implied by this fact via looking at the truncates in H of vectors in $\hat{H}$. The norms of these are characterized by the following theorem.

<u>Theorem</u>: Let $\hat{x}$ be an element of an extension space $(\hat{H},\hat{E})$ of a resolution space (H,E). Then $\hat{x}$ is in H if and only if

$$\limsup_{c \text{ in } C} ||\hat{E}^c\hat{x}|| < \infty$$

for every co-final set $C \subset G$. Moreover, when $\hat{x}$ is in H the above lim-sup is equal to the norm of $\hat{x}$.

Proof: If $\hat{x}$ is in H then $\hat{E}^c\hat{x} = E^c\hat{x}$ and since E^c goes strongly to the identity on H we have

$$\limsup_{c \text{ in } C} ||\hat{E}^c\hat{x}|| = \limsup_{c \text{ in } C} ||E^c\hat{x}|| = ||\hat{x}|| < \infty$$

On the other hand, if the lim-sup is finite, say bounded by M, then since the ball of radius M in H is weakly compact there exists a subnet; $\hat{E}^s\hat{x}$, s in $S \subset C$; of the $\hat{E}^c\hat{x}$ which converges weakly in H, say to <u>x</u> (YS-1).

Thus since E^r, r in S, is bounded the subnet $E^r\hat{E}^s\hat{x}$ converged weakly to $E^r\underline{x}$. Finally, for $r \leq s$ $E^r\hat{E}^s\hat{x} = \hat{E}^r\hat{x}$ is independent of s; hence, the existence of the limit implies that

$$\hat{E}^r\hat{x} = E^r\underline{x}$$

which by the uniqueness condition in hypothesis iv) for the extension space implies that $x = \underline{x}$ is in H as was to be shown.

Consistent with the theorem we may extend the norm from H to $\hat{H}$ by letting it be infinity if $\hat{x}$ is in $\hat{H}\backslash H$. Similarly, we may define a possibly infinite sup norm for the operators on H in the usual manner. The theorem, of course, assures that both of these norms will coincide with the usual Hilbert spaces norms when they are defined.

3. Causal Extensions

All of the algebraic aspects of causality carry over from a resolution space to its extension space without modification if we say that an operator, $\hat{T}$, on an extension space, $(\hat{H},\hat{E})$ is causal if whenever $\hat{E}^t\hat{x} = \hat{E}^t\hat{y}$ then $\hat{E}^t\hat{T}\hat{x} = \hat{E}^t\hat{T}\hat{y}$ for all $\hat{x}$, $\hat{y}$ in $\hat{H}$ and T in G. The same arguments as were used in the normed resolution space case (1.B.1) yield:

Theorem: For a linear operator, $\hat{T}$, on an extension space, $(\hat{H},\hat{E})$, the following are equivalent.

i) $\hat{T}$ is causal.

ii) $\hat{E}^t\hat{T} = \hat{E}^t\hat{T}\hat{E}^t$ for all t in G.

iii) If $\hat{E}^t\hat{x} = 0$ then $\hat{E}^t\hat{T}\hat{x} = 0$ for all $\hat{x}$ in $\hat{H}$ and t in G.

iv) $\hat{T}\hat{E}_t = \hat{E}_t\hat{T}\hat{E}_t$ for all t in G.

v) $\hat{E}^t\hat{T}\hat{E}_t = 0$ for all t in G.

In many respects the asymetry of the extension space makes it a more natural setting for the asymmetrical causal operators than the resolution spaces in which they have thus far been studied. In fact:

<u>Theorem</u>: Let $(\hat{H},\hat{E})$ be an extension space of (H,E) and let T be a causal operator on (H,E). Then T has a unique causal extension, $\hat{T}$, on $(\hat{H},\hat{E})$.

Proof: For any $\hat{x}$ in $\hat{H}$ we let

$$y^t = E^t T \hat{E}^t \hat{x}$$

where the right side of the equality is well defined since the range of $\hat{E}^t$ is in H. Now since T is causal and $\hat{E}$ extends E, if $s \leq t$ we have

$$E^s y^t = E^s E^t T \hat{E}^t \hat{x} = E^s T \hat{E}^t \hat{x} = E^s T E^s \hat{E}^t \hat{x} = E^s T \hat{E}^s \hat{E}^t \hat{x}$$

$$= E^s T \hat{E}^s \hat{x} = y^s$$

hence the definition of the extension space assures that there exists a unique $\hat{y}$ in $\hat{H}$ such that

$$\hat{E}^t \hat{y} = y^t$$

for all t in G and we may define $\hat{T}$ by the equality

$$\hat{y} = \hat{T}\hat{x}$$

If x is in H

$$\hat{E}^t \hat{T} x = \hat{E}^t \hat{y} = y^t = E^t T \hat{E}^t x = E^t T E^t x = E^t T x$$

hence the uniqueness condition for the extension space assures that $Tx = \hat{T}x$ and hence that $\hat{T}$ is an extension of T. Similarly, for any $\hat{x}$ in H

$$\hat{E}^t \hat{T} \hat{x} = E^t T \hat{E}^t \hat{x} = \hat{E}^t \hat{T} \hat{E}^t \hat{x}$$

since $\hat{T}$ and $\hat{E}$ extend T and E, respectively. $\hat{T}$ is thus causal.

Finally, if $\tilde{T}$ is another causal extension of T in $(\hat{H},\hat{E})$ then for any $\hat{x}$ in $\hat{H}$ we have

$$\hat{E}^t\hat{T}\hat{x} = \hat{E}^t\hat{T}\hat{E}^t\hat{x} = \hat{E}^t T\hat{E}^t\hat{x} = \hat{E}^t\tilde{T}\hat{E}^t\hat{x} = \hat{E}^t\tilde{T}\hat{x}$$

whence the uniqueness condition for the extension space implies that $\hat{T}\hat{x} = \tilde{T}\hat{x}$ showing that the causal extension is unique and completing the proof of the theorem.

4. Well-Posed Feedback Systems

Consistent with the causal extension theorem the operators appearing in the equations defining a feedback system may be extended to operators defined on an extension space $(\hat{H},\hat{E})$ of (H,E) and we may attempt to find solutions to the equations in $(\hat{H},\hat{E})$ rather than (H,E). We say that a feedback system is well-posed on $(\hat{H},\hat{E})$ if $(\hat{I} - \hat{K}\hat{F})^{-1}$ exists and is causal on $(\hat{H},\hat{E})$. We say that a feedback system is well-posed if it is well-posed on some extension space. Here $(\hat{I} - \hat{K}\hat{F})$ is the extension of (I - KF) to $(\hat{H},\hat{E})$ and similarly for $(\hat{I} - \hat{F}\hat{K})$, $\hat{F}$, $\hat{K}$, $\hat{K}\hat{F}$ and $\hat{F}\hat{K}$.

Although many systems are not well-posed in (H,E) they may be well-posed in an appropriate extension space. In particular, this includes most of the differential and difference systems commonly studied. For instance, if KF is given by

$$(KFy)(k) = \sum_{i=0}^{k-1} g(k,i)y(i)$$

on the resolution space $L_2(Z,R,m^+)$ with its usual spectral measure, then the operator equality

$$z = (I - KF)y$$

is characterized by

$$z(k) = y(k) - \sum_{i=0}^{k-1} g(k,i)y(i)$$

hence given z one may compute y via

$$y(0) = z(0)$$

$$y(1) = z(1) + g(1,0)y(0)$$

and repeating the process recursively we obtain

$$y(k) = z(k) + \sum_{i=0}^{k-1} g(k,i)y(i)$$

Now in general there is no reason to expect the y obtained by this process to satisfy a boundedness condition but it is a well defined element in the space of all one-sided sequences for any z in this vector space. As such $(\hat{I} - \hat{K}\hat{F})^{-1}$ exists in this extension space (with its usual algebraic spectral measure) and it is clearly causal via the recursive nature of its construction. A feedback system with its open loop gain so defined is thus well-posed.

Similarly, if the open loop gain is characterized by an integral operator or (linear, constant coefficient, ordinary) differential equation on $(L_2(R,R,m)$ standard integral and differential equation techniques may be invoked to verify the well-posedness of the feedback system on the space of functions for which

$$\int_{(-\infty,t)} ||f(q)||^2 dm(q) < \infty$$

B. Stability

1. Stable Operators

Conceptually, one would like the concept of the stability of an operator to correspond to the requirement that bounded inputs yield bounded outputs. Unfortunately, since we deal with operators defined on an extension space the inputs and outputs may not be normed and such an approach cannot be applied directly. Fortunately, however, the truncates of a vector in an extension space are normed and completely characterize the vector. As such, we may require that inputs with bounded truncates yield outputs with bounded truncates. In particular, if $\hat{T}$ is an operator on an extention space $(\hat{H},\hat{E})$ of (H,E) we say that $\hat{T}$ is stable if there exists a real constant M such that

$$||\hat{E}^t\hat{T}\hat{x}|| \leq M||\hat{E}^t\hat{x}||$$

for all $\hat{x}$ in $\hat{H}$ and t in G. Since the truncates of elements in $\hat{H}$ are in H these norms are assured to exist and be finite.

Although the definition of stability is in terms of the extension space one can often study the stability properties of an operator on an extension space via its restriction to the underlying Hilbert space (WL-1, WL-3, SA-10). In particular:

<u>Theorem</u>: Let $\hat{T}$ be an operator on an extension space $(\hat{H},\hat{E})$ of (H,E) and let T be the restriction of $\hat{T}$ to H. Then $\hat{T}$ is stable if and only if $\hat{T}$ is causal and T is a bounded operator mapping H into H.

Proof: If $\hat{T}$ is stable and x is in H there exists an M such that

$$||\hat{E}^t\hat{T}x|| \leq ||\hat{E}^tx||M = ||E^tx||M \leq ||x||M$$

Thus

$$\limsup_{t \text{ in } G} ||\hat{E}^t\hat{T}x|| \leq ||x||M$$

showing that $\hat{T}x$ is in H and that its norm is less than or equal to $||x||M$ hence T maps H into H and has norm less than or equal to M. Now if $\hat{T}$ is not causal there must exist an $\hat{x}$ in $\hat{H}$ such that $\hat{E}^t\hat{x} = 0$ but $\hat{E}^t\hat{T}\hat{x} \neq 0$ for some t. Hence for any M

$$M||\hat{E}^t\hat{x}|| - ||\hat{E}^t\hat{T}\hat{x}|| = -||\hat{E}^t\hat{T}\hat{x}|| < 0$$

where the strict inequality is due to the fact that $\hat{E}^t\hat{T}\hat{x} \neq 0$. As such

$$||\hat{E}^t\hat{T}\hat{x}|| > M||\hat{E}^t\hat{x}||$$

for all M and the operator T is not stable if it is not causal.

Conversely, if T maps H into H, is bounded and $\hat{T}$ is causal, then for any $\hat{x}$ in $\hat{H}$ and t in G let $x^t = \hat{E}^t\hat{x}$.

Now since T maps H into H and is bounded there exists an M independent of $\hat{x}$ such that

$$M||\hat{E}^t x||^2 = M||x^t||^2 \geq ||Tx^t||^2 = ||E^tTx^t||^2 + ||E_tTx^t||^2$$

$$\geq ||E^tTx^t||^2 = ||\hat{E}^t\hat{T}\hat{E}^t x||^2 = ||\hat{E}^t\hat{T}x||^2$$

which is the required inequality to show that $\hat{T}$ is stable. Note that the last equality follows from the causality of $\hat{T}$ and the second to the last from the fact that $\hat{T}$ and $\hat{E}$ extend T and E respectively. The proof is thus complete.

2. Stable Feedback Systems

We define the stability of a feedback system via the stability of the operators which it defines between the independent and dependent variables in any extension spaces where they are defined. As such, we say that a well-posed feedback system is stable if and only if the operator $(\hat{I} - \hat{K}\hat{F})^{-1}$ is stable on every extension space where it is defined. Note that since K is causal and bounded, $\hat{K}$ is assured to be stable and hence stability of the feedback system implies stability of the operator

$$(\hat{I} - \hat{K}\hat{F})^{-1}\hat{K}$$

Similarly, the stability of $\hat{F}$ together with our theorem relating $(\hat{I} - \hat{F}\hat{K})^{-1}$ to $(\hat{I} - \hat{K}\hat{F})^{-1}$ imply the stability of

$$(\hat{I} - \hat{F}\hat{K})^{-1}$$

and

$$\hat{F}(\hat{I} - \hat{F}\hat{K})^{-1}$$

for a stable feedback system. As such, for a stable feedback system all of the operators in the equations relating y and e to u and d are stable.

Although the physical interpretation of stability is highly dependent on our assumption that the feedback system is well-posed in an extension space rather than the original resolution space, since all vectors in the resolution space are, in a sense, bounded; the characterization of the stability of a feedback system can be carried out entirely in the underlying resolution space without ever explicitly constructing the extension space. Of course, this is an extremely powerful tool since it allows one to invoke Hilbert space, rather than vector space, methods in the characterization of stability (WL-3).

Theorem: A feedback system is well-posed and stable if and only if the following are satisfied.

i) $(I - KF)^{-1}$ exists on H.

ii) $(I - KF)^{-1}$ is bounded.

iii) $(I - KF)^{-1}$ is causal on (H,E).

Proof: Since the feedback system is well-posed there must exist an extension space $(\hat{H},\hat{E})$ on which $(\hat{I} - \hat{K}\hat{F})^{-1}$ exists and since the system is stable the operator $(\hat{I} - \hat{K}\hat{F})^{-1}$ is stable. The characterization theorem thus implies that the restriction of $(\hat{I} - \hat{K}\hat{F})^{-1}$ to H is a bounded operator mapping H into H. As such, the restriction of $(\hat{I} - \hat{K}\hat{F})^{-1}$ to H is a bounded inverse of (I - KF) on H, and since the system is well-posed $(\hat{I} - \hat{K}\hat{F})^{-1}$ is causal hence so is its restriction to H.

Conversely, if $(I - KF)^{-1}$ exists on H and is bounded and causal then for any extension space $(\hat{H},\hat{E})$ on which the feedback system is well-posed $(\hat{I} - \hat{K}\hat{F})^{-1}$ exists and is causal. Hence it is a causal extension of $(I - KF)^{-1}$ and by the uniqueness of the causal extension

$$||\hat{E}^t(\hat{I} - \hat{K}\hat{F})^{-1}\hat{x}|| = ||E^t(I - KF)^{-1}\hat{E}^t\hat{x}|| \leq M||\hat{E}^t\hat{x}||$$

for any $\hat{x}$ in $\hat{H}$ (2.A.3). Here M is the norm of $(I - KF)^{-1}$ and the above inequality verifies the stability of the feedback system on $(\hat{H},\hat{E})$ thereby completing the proof.

Note that the theorem essentially says that if a well-posed system is stable on some extension space it is also well-posed (and stable) on the underlying resolution space; hence one need never deal with the extension space when studying stable feedback systems. As such, extension spaces, though physically necessary for the definition of stability, do not appear in the sequel wherein stability is always studied via the theorem.

Finally, since the existence, boundedness and causality of $(I - KF)^{-1}$

and $(I - FK)^{-1}$ are equivalent the above theorem can be restated in terms of FK, as can the stability theorems to follow.

3. A Monotonicity Theorem

The preceding theorem yields an immediate tool for characterizing the stability of a feedback system in terms of the inverse return difference. However, since inversion often complicates the description of an operator it is desirable to characterize the stability of a feedback system directly in terms of the open loop gain KF rather than the inverse return difference. Our first such theorem (WL-1, DA-1) of this type is:

Theorem: For a feedback system defined on (H,E) let the open loop gain have a negative hermitian part (i.e., $F^*K^* + KF \leq 0$). Then the feedback system is well-posed and stable.

Proof: If KF has negative hermitian part then the return difference $(I - KF)$ has a positive definite hermitian part thereby guaranteeing the existence and boundedness of $(I - KF)^{-1}$. Moreover, since $(I - FK)$ has a strictly positive hermitian part

$$\mathrm{Re}\langle x,(I - KF)x\rangle > 0$$

for $x \neq 0$. As such, the causal invertibility theorem (1.C.4) assures that $(I - KF)^{-1}$ is causal. The preceding theorem thus assures that the feedback system is well-posed and stable; hence the proof is complete.

4. A Small Gain Theorem

Another criterion for the stability of a feedback system in terms of KF requires that it have small norm (DA-1, WL-1, DS-1, DE-1).

Theorem: For a feedback system defined on (H,E) let $||(KF)^N|| < 1$ for some positive integer N. Then the feedback system is well-posed and stable.

Proof: Let

$$||(KF)^N|| = c < 1$$

and let

$$\underset{1 \leq i < N}{\text{maximum}} ||(KF)^i|| = M$$

Then for any integer n we let $n = kN + m$ where $1 \leq m < N$ and have

$$||(KF)^n|| = ||(KF)^{kN}(KF)^m|| \leq ||(KF)^N||^k||(KF)^m||$$

$$\leq Mc^k$$

As such the Neumann series

$$\sum_{i=0}^{\infty} (KF)^i$$

converges and hence $(I - KF)^{-1}$ exists on H (since it is equal to the limit of the Neumann series when it converges), is bounded and causal (since the Neumann series yields a representation of $(I - KF)^{-1}$ as the limit of a sequence of causal operators (1.C.1)). The conditions of the stability theorem are thus satisfied and hence the system is well-posed and stable as was to be shown.

5. A Strict Causality Condition

Rather than requiring that the norm of the entire open loop gain be small as in the previous theorem it suffices to require merely that the memoryless part, if it exists, of the operator be small (DE-1, DE-3, DE-4, DE-5).

Theorem: For a feedback system defined on (H,E) with open loop gain KF let

$$C\int_G dE(t)KFdE(t)$$

exist and have norm less than one. Then the feedback system is well-posed and stable.

Proof: If

$$||C\int_G dE(t)KFdE(t)|| < 1$$

there exists a partition $-\infty = t_o < t_1 < t_2 < \ldots < t_n = \infty$ of G such that

$$||\sum_{i=1}^{n} [E^{t_i}-E^{t_{i-1}}]KF[E^{t_i}-E^{t_{i-1}}]||^2 < 1$$

and thus

$$||[E^{t_i}-E^{t_{i-1}}]KF[E^{t_i}-E^{t_{i-1}}]||^2 < 1$$

for all i. Now let H_i be the range of $[E^{t_i}-E^{t_{i-1}}]$ and represent H as the orthogonal sum of the H_i. Similarly, the operator (I - KF) may be represented as the n by n matrix of operators

$$(I - KF)_{ij} = [E^{t_i}-E^{t_{i-1}}](I - KF)[E^{t_j}-E^{t_{j-1}}]$$

where $(I - KF)_{ij}$ is a causal operator mapping H_j to H_i. Now since KF is causal

$$(I - KF)_{ij} = \begin{cases} 0 & ; i < j \\ I_i - (KF)_{ii} & ; i = j \\ (KF)_{ij} & ; i > j \end{cases}$$

where I_i is the identity on H_i and

$$(KF)_{ij} = [E^{t_i}-E^{t_{i-1}}]KF[E^{t_j}-E^{t_{j-1}}]$$

(I - KF) thus has the matrix representation

$$(I - KF) = \begin{bmatrix} I_1-(KF)_{11} & & & \\ (KF)_{21} & I_2-(KF)_{22} & \llcorner 0 & \\ (FK)_{21} & (KF)_{32} & & \\ \vdots & & & \\ (FK)_{n1} & (KF)_{n2} & \cdots & I_n-(FK)_{nn} \end{bmatrix}$$

and therefore its inverse may be computed by the usual formula for the inverse of a triangular matrix which involves the $(KF)_{ij}$ and the inverses of the $I_i-(KF)_{ii}$. Now the former are causal and bounded since (KF) is causal and bounded and the latter exists and are causal and bounded via the small gain theorem (2.B.4) since the $(KF)_{ii}$ are causal, bounded and of norm less than one. As such, it is possible to write $(I - FK)^{-1}$ explicitly in terms of a finite number of causal bounded operators thereby assuring that it exists, is causal and bounded. Hence the feedback system is well-posed and stable as required.

An equivalent formulation of the preceding small memoryless part theorem is:

<u>Theorem</u>: For a feedback system defined on (H,E) let the open loop gain be decomposed as

$$(KF) = (KF)_C + (KF)_M$$

where $(KF)_C$ is strictly causal and $(KF)_M$ is memoryless with norm less than one. Then the feedback system is well-posed and stable.

Proof: By the decomposition theorem (1.F.1)

$$(KF)_M = C\int_G dE(t)FKdE(t)$$

hence this follows from the preceding result.

Theorem: For a feedback system defined on (H,E) let the open loop gain KF be a strictly causal operator. Then the feedback system is well-posed and stable.

Proof: This is just the special case of the preceding theorem based on operator decomposition where the memoryless term is zero.

We note that all three of the preceding results are strongly predicated on the assumption that the diagonal integral (operator decomposition, strictly causal operator) is defined with the limit taken in the uniform operator topology (DE-1). In fact, the results fail in the strong operator topology where there are many strongly strictly causal differential operator open loop gains which yield unstable feedback systems.

6. A Necessary and Sufficient Condition

It is possibly surprising that it is possible to give a necessary and sufficient condition for the stability of a feedback system wholly in terms of the open loop gain (DA-1, DA-3). The condition, however, is not readily tested.

Theorem: Let a feedback system be defined on (H,E). Then a necessary and sufficient condition for it to be well-posed and stable is that there exist an $\varepsilon > 0$ such that the open loop gain KF satisfies

$$||(I-E^t(KF)-E_t(KF)^*)x||^2 \geq \varepsilon||x||^2$$

for all x in H and t.

Proof: The necessary and sufficient condition (1.C.3) for the return difference, which is a bounded causal operator, to have a bounded causal inverse is that there exist an $\varepsilon > 0$ such that

$$||E^t(I-KF)x||^2 + ||E_t(I-KF)^*x||^2 \geq \varepsilon||x||^2$$

for all x and t. Now upon invoking the fact that the range of E^t and E_t are orthogonal we have

$$\begin{aligned}\varepsilon||x||^2 &\leq ||E^t(I-KF)x||^2 + ||E_t(I-KF)^*x||^2 \\ &= ||E^t(I-KF)x + E_t(I-KF)^*x||^2 \\ &= ||(E^t-E^t(KF)+E_t-E_t(KF)^*)x||^2 \\ &= ||(I-E^t(KF)-E_t(KF)^*)x||^2\end{aligned}$$

which was to be shown.

Note that the theorem can be restated as the requirement that the operators

$$(I-E^t(KF)-E_t(KF)^*)$$

have uniformly bounded (left) inverses for all t; hence the theorem essentially translates the determination of whether or not an operator has a bounded causal inverse into an equivalent problem of whether or not a family of operators all have bounded inverses.

C. Sensitivity

1. Open Loop Systems

In most systems the plant K is given a-priori but the feedback is artificially added. In fact, its physical manifestation is often no more than a subroutine in a computer program. As such, rather than working with a feedback system one can just as easily deal with an open loop system such as shown below for which no questions of inversibility or

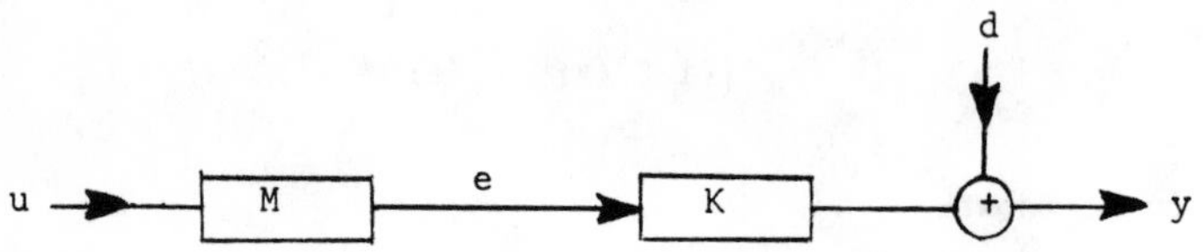

Figure: Configuration for Open Loop System

stability arise. Here the system is characterized by the equations

$$y = Ke + d$$

and

$$e = Mu$$

and the operators defined by the system are

$$y = KMu + d$$

and

$$e = Mu$$

which always exist and are stable for a causal plant K and compensator M.

Since the plant equations for the feedback and open loop systems are the same and the designer usually has control of the remainder of the system one can often build equivalent systems via either approach (PR-4, AN-2). Since d is usually interpreted as a disturbance for which the designer has no control we define equivalence only with respect to the mappings between u and y and u and e with d taken to be zero. That is, we say that a feedback and an open loop system are <u>equivalent</u> if $e_f = e_o$ and $y_f = y_o$ whenever $d_f = d_o = 0$ and $u_f = u_o = u$ for all u in J. Here the "f" subscripts denote variables for the feedback system and the "o" subscripts denote variables for the open loop system.

<u>Theorem</u>: Let a feedback system and an open loop system have the same plant $K_f = K_o = K$ and be equivalent. Then

$$I - M + FKM = 0$$

Proof: If $e_f = e_o$ for d = 0 and all u then

$$0 = e_f - e_o = u_f + Fy_f - Mu_o = u_f + FKe_f - Mu_o = u_f + FKe_o - Mu_o$$
$$= u_f + FKMu_o - Mu_o = (I-M+FKM)u$$

for all $u = u_f = u_o$ which was to be shown.

2. Sensitivity Comparison

Given equivalent open loop and stable feedback systems all system variables are identical and hence the two systems are indistinguishable. The main factor, however, which usually leads one to deal with feedback rather than open loop systems is their sensitivity to perturbations in the plant K. That is, if the actual plant K differs from the nominal plant K' for which the systems were designed the errors resulting in the system output

$$\delta_f = y_f' - y_f$$

and

$$\delta_o = y'_o - y_o$$

tend to be greater in the open loop system than the feedback system. Here the primed quantities denote the nominal system outputs resulting from the plant K' and the unprimed quantities denote the actual outputs resulting from the plant K. Although it is quite hard to determine the individual δ's it is, possibly surprisingly, possible to compute their ratio and hence compare the two systems without actually determining their solutions. For this purpose we tabulate the equations characterizing the two systems in their nominal and actual forms as follows. Of course, $d_f = d_o = d'_f = d'_o = 0$ and $u_f = u_o = u'_f = u'_o = u$ since the inputs are independent of the systems and/or the plant. Denoting feedback variables by an "f" subscript, open loop variables by an "o" subscript, nominal variables by a prime and actual variables without a prime we have:

$$y_f = Ke_f$$

and

$$e_f = Fy_f + u$$

for the actual feedback system;

$$y'_f = K'e'_f$$

and

$$e'_f = Fy'_f + u$$

for the nominal feedback system;

$$y_o = Ke_o$$

and

$$e_o = Mu$$

for the actual open loop system; and

$$y_o' = K'e_o'$$

$$e_o' = Mu$$

for the nominal open loop system. Note that $e_o' = e_o$ in this case since we assume no perturbation in M nor F. This is due to an assumption that the feedback and compensation operators are "in the hands of the system designer" and hence can be constructed with sufficient accuracy to render any perturbation in their behavior negligible. On the other hand, the plant K is usually given a-priori and may tend to have large perturbations due to environmental effects, time-variations or simply bad modeling.

Although the complexity of our feedback system is such as to render the actual computation of the error δ_f due to perturbations in the plant an impossibility it is possible, under the preceding assumptions to compute the relationship between δ_f and δ_o exactly and independently of the source of the perturbation in the plant (PR-4, AN-2).

<u>Theorem</u>: Let a well-posed stable feedback system with plant K (with nominal value K') and feedback F be nominally equivalent to an open loop system with the same plant and compensator M. Then

$$\delta_f = (I - KF)^{-1}\delta_o$$

for all values of the system input u.

Proof: By definition $\delta_f = y_f' - y_f$ and $\delta_o = y_o' - y_o$ hence since the equivalence of the two systems implies that $y_o' = y_f'$ we have

$$\delta_o = \delta_f + (y_f - y_o)$$

and

$$y_f - y_o = KFy_f + Ku - KMu = K(I-M(u + KFy_f$$

Now upon substitution of the result of the equivalence theorem for the nominal system values

$$KFy_f' = KFK'e_f' = KFKe_o' = K(FK'M)u = -K(I-M)u$$

Combining these results thus yields

$$\begin{aligned}\delta_o &= \delta_f + (y_f-y_o) = \delta_f + K(I-M)u + KFy_f \\ &= \delta_f - KFy_f' + KFy_f = \delta_f - KF(y_f'-y_f) = \delta_f - KF\delta_f \\ &= (I-KF)\delta_f\end{aligned}$$

Finally, since we have assumed that the feedback system is stable $(I-KF)^{-1}$ exists and is causal; hence we may write

$$\delta_f = (I-KF)^{-1}\delta_o$$

independently of u and our proof is complete.

3. Sensitivity Reduction Criteria

Consistent with the preceding theorem if $||(I-KF)^{-1}||$ is less than or equal to one we are assured that the feedback system is at least as insensitive to perturbations in the plant as the open loop system. Conversely, if $||(I-KF)||$ is less than or equal to one the open loop system is superior, whereas if neither condition is satisfied one cannot make an immediate statement about the relative sensitivities of the two systems. We note that since $(I-KF)$ and $(I-KF)^{-1}$ are assured to be causal our conditions for sensitivity reduction are equivalent to the requirement that operators

(I-KF) and $(I-KF)^{-1}$ be S-passive. That is, an operator T on a resolution space (H,E) is S-passive if

$$||E^t Tx|| \leq ||E^t x||$$

for all x in H and t in G. This is precisely the stability criterion for an operator with M taken to be one (2.B.1), and by adopting the corresponding theorem for that case we may verify our contention that the sensitivity criterion for $(I-KF)^{-1}$ is equivalent to S-passivity.

Theorem: Let T be an operator on a resolution space (H,E). Then T is S-passive if and only if it is causal and contractive.

The source of the term S-passivity is due to the fact that the S-passivity condition is precisely the condition used to define the passivity of an electric network when T is given the physical interpretation of a network scattering operator.

4. A Generalized Sensitivity Theorem

On occasion rather than dealing with a plant and feedback operator only, we also like to include a compensator in the feedback system as shown below. Here the equations for this compensated feedback system are

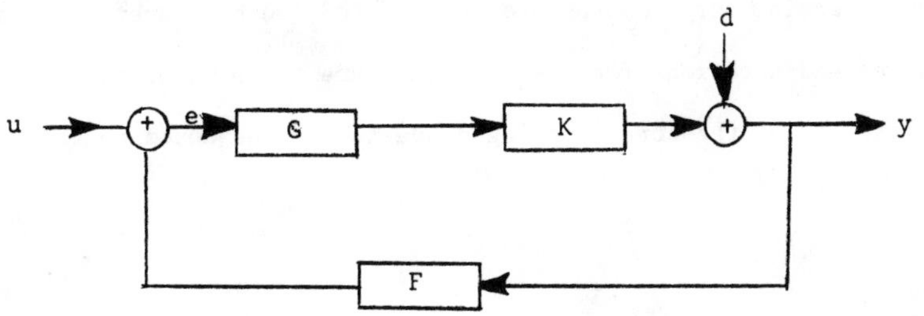

Figure: Feedback System with Compensator

$$y = KGe + d$$

and

$$e = Fy$$

Of course, for most purposes (for instance, well-posedness and stability) one can simply treat KG as the plant in the compensated feedback system and apply the standard results already derived for the regular feedback system to the compensated case. For the sensitivity problem, however, this is not the case since we wish to consider only perturbations in K and not KG. It is, however, possible to formulate a theorem analogous to the preceding in this case wherein one applies essentially the same argument with the obvious generalizations (AN-2).

Theorem: Let a well-posed stable compensated feedback system with plant K (with nominal value K'), feedback F and compensator G be nominally equivalent to an open loop system with the same plant and compensator M. Then

$$\delta_{cf} = (I-KGF)^{-1}\delta_o$$

for all values of the system input u.

Here the subscript "cf" denotes the compensated feedback system and all definitions and notations for these systems are taken to be the "obvious" generalizations of the corresponding concepts for the uncompensated case.

D. Optimal Controllers

1. The Control Problems

In general one can construct feedback systems where $(I-KF)^{-1}$ has any norm hence no a-priori statements can be made about the superiority of the feedback system over the open loop system. If, however, the feedback and open loop systems are designed to satisfy a specific goal it is often possible to make a definitive statement. In particular, if the systems are designed to solve an optimal control problem this is the case (PR-3, PR-4).

The feedback control problem may be formalized as: Given a causal operator K on a resolution space (H,E) find an operator F on (H,E) such that

i) F is causal.

ii) The feedback system with open loop gain KF is well-posed and stable.

iii) The functional

$$J = ||y||^2 + ||e||^2$$

is minimized under the constraints

$$y = Ke + d$$

and

$$e = Fy$$

for all d in a given subspace D of H.

Note that the constraints of the feedback control problem are precisely those for a feedback system with $u = 0$.

An alternate control problem corresponds to the case where one is dealing with an open loop system rather than a feedback system. The open loop control problem may be formalized as: Given a causal operator K on a resolution space (H,E) find an operator M on (H,E) such that:

i) M is causal.

ii) The functional

$$J = ||y||^2 + ||e||^2$$

is minimized under the constraints

$$y + Ke + d$$

and

$$e = Md$$

for all d in a given subspace D of H.

Note that we need not consider stability in the open loop problem since there is no feedback involved in its formulation though for "physical realizability" M as well as F must be causal. Here the constraints for the open loop control problem are precisely those of an open loop system with the input u taken to be d.

Physically, if we interpret d as a disturbance on the system both control problems amount to choosing an input for the plant which causes the plant output to cancel d by making the norm of y (equal to d plus the plant output) small with the added constraint that the norm of e (the energy or fuel supplied to the plant) does not become large. In one case the control e is obtained as a causal function of the system output y and in the other case it is obtained as a causal function of the disturbance, d, itself. The later approach is somewhat open to question on physical grounds since it is often not possible to measure d directly except via its manifestations in y. Mathematically, both problems are, however, quite viable and essentially equivalent.

Since we require the solution of the optimal control problems to hold independently of d in D they are not, in general, solvable. There are, however, a number of special cases, which include the most common practical control problems, wherein the special properties of K and/or D allow solutions to be obtained. Our purpose, however, is not the solution of the control problems. Rather, we will simply verify the "equivalence" of the two control problems and determine their relative sensitivities under the assumption that solutions F and M have been given. In the sequel we denote by $J_o(d)$ and $J_f(d)$ the, d dependent, values of the functional obtained for the given solutions of the open loop and feedback control problems, respectively.

2. Equivalence Theorems

Although the physical viability of the open loop control problem (as well as its practical implementation) is open to question the two control problems are essentially equivalent, their difference lying in the fact that there is no stability condition on the open loop problem whence in certain "pathological" cases the open loop problem may have a "better" solution than the feedback problem.

Lemma: Let $J_o(d)$ be the value of $||y||^2 + ||e||^2$ achieved for a solution of the open loop control problem for each d in D, and let $J_f(d)$ be the value of $||y||^2 + ||e||^2$ achieved for a solution of the feedback control problem for each d in D (both with a fixed K, (H,E) and D). Then

$$J_o(d) \le J_f(d)$$

for all d in D.

Proof: Let F be a solution for the feedback control problem and let

$$M = (I-FK)^{-1}F$$

Now since F solves the feedback control problem it is causal and the feedback system with open loop gain KF is well-posed and stable. That is $(I-KF)^{-1}$ exists, is bounded and causal; and hence $(I-FK)^{-1}$ also exists, is bounded and causal. M thus exists, bounded and is causal. Now since F solves the feedback control problem

$$J_f(d) = ||\underline{y}||^2 + ||\underline{e}||^2$$

for all d in D where

$$\underline{y} = K\underline{e} + d$$

and

$$\underline{e} = F\underline{y}$$

Thus

$$\underline{e} = F(K\underline{e}+d) = FK\underline{e} + Fd$$

or equivalently

$$\underline{e} = (I-FK)^{-1}Fd = Md$$

$\underline{e}$ and $\underline{y}$ thus satisfy the constraints for the open loop control problem with M causal and thus

$$J_o(d) \leq ||\underline{y}||^2 + ||\underline{e}||^2 = J_f(d)$$

for all d in D (since the $J_o(d)$ is the minimum over all e and y which satisfy the above equalities for some causal M and hence is less than or equal to the value of the functional for this special case). The proof is thus complete.

With the aid of the above lemma we may now prove:

Theorem: Let K, (H,E) and D be given and let M be a solution of the open loop control problem for which $(I+KM)^{-1}$ exists and is causal. Then $J_o(d) = J_f(d)$ for all d in D and

$$F = M(I+KM)^{-1}$$

is a solution for the feedback control problem.

Proof: Since M is a solution to the open loop control problem it is causal and $(I+KM)^{-1}$ exists and is causal (by hypothesis), hence F exists and is causal. Now

$$(I-KF)^{-1} = (I-KM(I-KM)^{-1})^{-1}$$

$$= (((I+KM)-KM)(I+KM)^{-1})^{-1} = (I+KM)$$

exists, is bounded and causal; hence the feedback system with open loop gain KF is well-posed and stable. Now, since M solves the open loop control problem

$$J_o(d) = ||\underline{y}||^2 + ||\underline{e}||^2$$

for all d in D where

$$\underline{y} = K\underline{e} + d$$

and

$$\underline{e} = Md$$

Thus

$$\underline{e} = Md = M(\underline{y}-K\underline{e}) = M\underline{y} = MK\underline{e}$$

or equivalently

$$\underline{e} = (I+MK)^{-1}M\underline{y} = M(I+KM)^{-1}\underline{y} = F\underline{y}$$

$\underline{e}$ and $\underline{y}$ thus satisfy the constraints for the feedback control problem with an F which is causal and renders the feedback system with open loop gain KF stable. Thus

$$J_f(d) \leq ||\underline{y}||^2 + ||\underline{e}||^2 = J_o(d) \leq J_f(d)$$

for all d in D (since $J_f(d)$ is the minimum over all e and y which satisfy the above equalities for some causal F which renders the feedback system with open loop gain KF well-posed and stable, and hence $J_f(d)$ is less than or equal to the value of the functional for this special case). Here the last inequality is a statement of the result of the lemma, and the combination of the two inequalities implies that

$$J_f(d) = ||\underline{y}||^2 + ||\underline{e}||^2$$

and hence that

$$F = M(I-KM)^{-1}$$

actually solves the feedback control problem.

In essence the feedback control problem has the extra constraint that the feedback system with open loop gain KF be stable which has no analog in the open loop case. Thus in general, $J_o(d) \leq J_f(d)$, but, as in the case of the theorem, if the solution to the open loop problem happens to satisfy a similar constraint (that $(I+KM)^{-1}$ exists, is bounded and causal which is equivalent to the requirement that the feedback system with open loop gain -KM be well-posed and stable) then $J_o(d) = J_f(d)$ and, in fact, the e and y which solve the two problems are identical.

Since the condition of the theorem that (I+KM) have a causal bounded inverse has no analog when one starts with a feedback control a partial converse to the preceding theorem can be obtained by assuming a-priori that $J_f(d) = J_o(d)$ rather than proving it.

Theorem: Let K, (H,E) and D be given and let F be a solution of the feedback control problem for which $J_o(d) = J_f(d)$ for all d in D. Then $(I-FK)^{-1}$ exists and

$$M = (I-FK)^{-1}F$$

is a solution to the open loop control problem.

The proof of the theorem follows from essentially the same argument used for the proof of the lemma with our a-priori assumption that $J_o(d) = J_f(d)$ assuring that the open loop controller so constructed is in fact optimal. The details will not be given.

3. A Sensitivity Theorem

Consistent with the fact that the two optimal control problems are nominally equivalent in the sense that they yield $J_o(d) = J_f(d)$ given the appropriate assumptions we would like to compare the systems via their sensitivity to plant perturbations. This can, however, not be achieved directly via the sensitivity theorem thus far derived since that is predicated on the assumption that d = 0 and the two systems have the same u whereas for the control problems we have $u = d \neq 0$ in the open loop case and u = 0 and $d \neq 0$ in the closed loop case. Fortunately, however, it is possible to make a transformation of systems wherein the control configurations are converted into configurations which satisfy the assumptions of the sensitivity theorem and have sensitivities which are identical to those of the original control configurations.

Lemma: Let an open loop system have compensator M, plant K, inputs $u_o = d_o = d$, plant input e_o, output y_o and deviation (of y_o with respect to perturbations in K) δ_o; and let a modified open loop system have the same plant and compensator, inputs $u_{mo} = d$ and $d_{mo} = 0$, plant input e_{mo}, output y_{mo} and deviation (of y_{mo} with respect to perturbations in K) δ_{mo}. Then

$$\delta_{mo} = \delta_o$$

$$e_{mo} = e_o$$

and

$$y_{mo} = y_o - d$$

Lemma: Let a feedback system have plant K, feedback F, inputs $u_f = 0$ and $d_f = d$, plant input e_f, output y_f and deviation (of y_f with respect to perturbations in K) δ_f; and let a modified compensated feedback system have the same plant, feedback $F_{mf} = I$, compensator $G_{mf} = F$, inputs $u_{mf} = d$ and $d_{mf} = 0$, plant input e_{mf}, output y_{mf} and deviation (of y_{mf} with respect to perturbations in K) δ_{mf}. Then

$$\delta_{mf} = \delta_f$$

$$e_{mf} = e_f$$

and

$$y_{mf} = y_f - d$$

The proofs of both lemmas are purely algebraic and will not be given.

We are now ready to prove our main theorem on the relative sensitivity of feedback and open loop optimal controllers (PR-3, PR-4).

Theorem: Let K', (H,E) and D be given and let M and F be optimal open loop and feedback controllers for K', (H,E) and D which are equivalent in the sense of the preceding theorems. Let the actual plant be

$$K = K' + L$$

where D is an invariant subspace of LM and let δ_f and δ_o be the deviations in the outputs of the feedback and open loop systems, respectively, due to the perturbation L for a fixed d. Then for any ε there exists a δ such that if $||L|| < \delta$ then

$$||\delta_f|| < ||(1+\varepsilon)\ \delta_o||$$

Proof: Consistent with the proofs of the equivalence theorems for the optimal controllers $e_f' = e_o'$ and $y_f' = y_o'$ for any d where the e's are the nominal values of the feedback and open loop plant inputs and the y's are the nominal values of the feedback and open loop system outputs, respectively. Now if we transform our feedback and open loop control systems to the modified systems of the lemma we have

$$u_{mo} = u_{mf} = d$$
$$e'_{mo} = e'_o = e'_f = e'_{mf}$$
$$y'_{mo} = y'_o - d = y'_f - d = y'_{mf}$$
$$\delta_{mo} = \delta_o$$

and

$$\delta_{mf} = \delta_f$$

The modified systems are thus equivalent and in the form for which the sensitivity theorem (for compensated feedback systems) applies. Recognizing that for the modified feedback system $F_m = I$, $G_m = F$ and K is the same as for the original system we have

$$\delta_f = \delta_{mf} = (I-K_m G_m F_m)^{-1}\delta_{mo} = (I-KF)^{-1}\delta_o$$

Finally we note that

$$\delta_o = K'Md + d - (KMd+d) = LMd$$

is in D via the invariant subspace assumption and that

$$J_o(d) = J_f(d) \leq ||d||^2$$

for all d in D since the trivial controller with M = F = 0 achieves this value for the functional. We thus have

$$||d||^2 \geq J_f(d) = ||y_f'||^2 + ||e_f'||^2 \geq ||y_f'||^2$$

and since

$$y_f' = K'e_f' + d = K'Fy_f' + d$$

implying that

$$y_f' = (I-K'F)^{-1}d$$

which shows that

$$||d||^2 \geq ||(I-K'F)^{-1}d||^2$$

for all d in D or equivalently that the norm of $(I-K'F)^{-1}$ restricted to D is less than or equal to one. Now since the mapping from K into $(I-KF)^{-1}$ is continuous (in the norm topology) on an open neighborhood of K' (since the inverse exists at K') for any ε there exists δ such that if $||L|| < \delta$ then

$$||(I-KF)^{-1}\Big|_D|| < 1 + \varepsilon$$

Now since δ_o is in D we have

$$||\delta_f|| = ||(I-KF)^{-1}\delta_o|| \leq ||(1+\varepsilon)\delta_o||$$

which was to be shown.

E. Problems and Discussion

1. Maximal Extension Spaces

The concept of well-posedness is somewhat ambiguous as we have thus far defined it since it is a property of the extension space used as well as the operator. In fact, this difficulty can be by-passed by always working with a maximal extension space. For this purpose we say that an extension space $(\hat{H},\hat{E})$ of (H,E) is greater than an extension space $(\tilde{H},\tilde{E})$ of (H,E) if $\hat{H} \supset \tilde{H}$ and $\hat{E}|_{\tilde{H}} = \tilde{E}$ with the maximal extension space being defined via this ordering in the obvious manner.

Problem: Show that every resolution space has a unique maximal extension space. (Use Zorn's Lemma).

Problem: Show that a feedback system is well-posed on maximal extension space $(\hat{H},\hat{E})$ of (H,E) if and only if it is well-posed on some extension space.

Consistent with the above results one can always deal with a maximal extension space of a resolution space thereby completely eliminating the choice of extension space from the well-posedness of a feedback system. Of course, our stability theorem (2.B.2) implies that we can always work in (H,E) itself if we desire to find stable well-posed feedback systems.

Problem: Determine the maximal extension space of $L_2(Z,R,m^+)$ with its usual resolution of the identity.

Problem: Determine the maximal extension space of $L_2(R,R,m)$ with its usual resolution of the identity.

Although an extension space is not normed the fact that the properties of a vector in the extension space are determined entirely by its truncation in (H,E) allows us to norm the operators on the extension space via

$$||\hat{T}|| = \sup_{\substack{t \\ ||\hat{x}|| \neq 0}} \frac{||\hat{E}^t \hat{T} \hat{x}||}{||\hat{E}^t \hat{x}||}$$

Problem: Show that the mapping which extends a causal operator in (H,E) to a causal operator in (H,E) is a Banach algebra isomorphism from the bounded causal operators on (H,E) onto the bounded (in the above norm) causal operators on $(\hat{H},\hat{E})$

2. Nonlinear Feedback Systems

As far as well-posedness and stability for feedback systems are concerned, much of the theory which we have thus far developed for linear systems goes over to the non-linear case (DA-1, WL-1, WL-3). In particular, we say that a feedback system is well-posed in $(\hat{H},\hat{E})$ if $(\hat{I}-\hat{K}\hat{F})^{-1}$ exists on $(\hat{H},\hat{E})$ and is same-input same-output causal and we say that the system is stable if the nonlinear operator $(\hat{I}-\hat{K}\hat{F})^{-1}$ is stable; in the sense that there exists an M such that

$$||\hat{E}^t(\hat{I}-\hat{K}\hat{F})^{-1}\hat{x}|| \leq M||\hat{E}^t\hat{x}||$$

for all x and t; on every extension space where it is well-posed.

Problem: Show that a same-input same-output causal nonlinear operator on an extension space is stable if and only if its restriction to the underlying Hilbert space is bounded.

Problem: Show that a well-posed (possibly) nonlinear feedback system is stable if and only if

i) $(I-KF)^{-1}$ exists on H.

ii) $(I-KF)^{-1}$ is bounded on H.

iii) $(I-KF)^{-1}$ is same-input same-output causal.

Consistent with the above and the study of nonlinear stability, like the linear case can be reduced to the study of a problem of causal invertibility on the underlying Hilbert space (WL-1). Surprisingly, most of our sufficient conditions for stability in the linear case generalize to the nonlinear case (DA-1, DA-2, DA-3, WL-1, WL-3, DE-1).

Problem: Show that if KF is a same-input same-output causal Lipschitz continuous operator with Lipschitz norm less than one, then the feedback system with open loop gain KF is stable. (Hint: use a contraction mapping theorem to show that (I-KF) is onto.)

The negativity condition also generalizes to the nonlinear case though, as in the previous problem where we assumed Lipschitz norm less than one rather than the usual norm, we must deal with a Lipschitz like negativity condition (DA-1, DA-2, DA-3). We therefore say that an operator, T, is monotone if

$$\langle x-y, T(x-y)\rangle \geq 0$$

Clearly, this reduces to the usual positivity condition for T in the linear case and we have:

Problem: Show that if KF is a same-input same-output causal Lipschitz continuous operator for which -KM is monotone, then the feedback system with open loop gain KF is stable.

In general the sufficient condition for stability formulated in terms of the diagonal integral of KF does not generalize to the nonlinear case. A partial generalization however is valid (DE-1).

Problem: Show that if KF = CN+L where C is a strictly causal linear operator, N is a memoryless nonlinear operator and L is a causal linear operator of norm less than one; then the feedback system with open loop gain KF is stable.

In the above stability condition CN is a weakly additive strictly causal operator; hence, it is natural to conjecture the following generalization.

Conjecture: Let KF = W+L where W is a weakly additive strictly causal operator and L is a causal linear operator of norm less than one. Then the feedback system with open loop gain KF is stable.

In general we believe that most linear stability results have generalizations to the weakly additive case though at this time the details have yet to be formalized.

3. Energy Constraints

The concepts of stability and S-passivity of an operator are closely related to the energy constraints one applies in linear network theory (SA-4, SA-9, WI-1, PR-1, PR-2, PR-4, PR-6). One additional such concept is S-losslessness wherein it is required that an operator be S-passive and also satisfy the equality

$$||Tx|| = ||x||$$

(i.e., be isometric) for all x in H.

Problem: Show that a linear operator is S-lossless if and only if it is causal and isometric.

Problem: Give an example of an isometric operator which is not S-lossless.

The passivity and losslessness concepts are predicated on the assumption that the norm has the physical interpretation of energy, which in network theory is indeed the case if T is given the physical interpretation of a scattering operator (SA-10), x that of an incident wave and Tx that of a reflected wave. If, however, one desires to work with the more classical (if less convenient) voltage and current variables (SA-10), energy takes the form of an inner product and we define <u>I-passivity</u> by the requirement that

$$\mathrm{Re}\langle E^t x, Tx\rangle \geq 0$$

for all x in H and t in G. Similarly, we say that T is <u>I-lossless</u> if it is I-passive and

$$\mathrm{Re}\langle x, Tx\rangle = 0$$

for all x in H.

<u>Problem</u>: Show that an operator is I-passive if and only if it is causal and has a positive semi-definite hermitian part.

<u>Problem</u>: Show that an operator is I-lossless if and only if it is causal and has a zero hermitian part.

<u>Problem</u>: Show that the inverse of an I-passive operator is I-passive if it exists.

Physically both modes of passivity and losslessness have the same physical interpretation though with respect to different sets of variables. As such, one would expect that a transformation of variables would convert an I-passive operator into an S-passive operator and vice-versa (SA-10). We therefore consider three operators, S, Z, and Y related via

$$S = (I+Y)^{-1}(I-Y) = (Z+I)^{-1}(Z-1)$$

$$Z = (I-S)^{-1}(I+S) = Y^{-1}$$

and

$$Y = (I+S)^{-1}(I-S) = Z^{-1}$$

Problem: For the above operators show that S is S-passive if and only if Y and Z are I-passive whenever the appropriate inverses exist.

Problem: For the above operators show that S is S-lossless if and only if Y and Z are I-lossless whenever the appropriate inverses exist.

If one gives the function

$$W(x,t) = ||E^t x||^2 = ||E^t Tx||^2$$

the physical interpretation of energy then one would expect to be able to divide W(x,t) into two terms corresponding to the energy dissipated and the energy stored (and potentially recoverable) from the system. For this purpose we denote the dissipated energy for an operator T by

$$D(x,t) = \inf_{\substack{y \\ E^t y = E^t x}} [||y||^2 - ||Ty||^2]$$

and the stored energy by

$$F(x,t) = W(x,t) - D(x,t)$$

Problem: For an S-passive operator T show that D(x,t) is a non-negative, non-decreasing function of t for each x.

Problem: Show that for an S-passive operator F(x,t) is a non-negative function of t which goes to zero at ∞.

The above concepts of dissipated and stored energy are predicated on the scattering variable interpretation of x and Tx. In the voltage current interpretation we write

$$\underline{W}(x,t) = \mathrm{Re}\langle E^t x, Tx\rangle$$

for our energy function.

Problem: Define dissipative and stored energy functions $\underline{D}(x,t)$ and $\underline{F}(x,t)$ in terms of the inner product and $\underline{W}(x,t)$ so as to assure that $\underline{D}(x,t)$ will be a non-negative, non-decreasing function of t for I-passive operators and $\underline{F}(x,t)$ will be a non-negative function of t which goes to zero as t goes to infinity for I-passive operators.

Problem: Characterize the D(x,t) and F(x,t) for S-lossless operators.

If physically S-passivity , i.e., $W(x,t) \geq 0$ for all x and t, represents a system with no internal energy sources, and hence the total energy entering the system at any time for an input is greater than or equal to zero, then the finite energy condition that there exists a real M such that

$$W(x,t) \geq M$$

for all x in H and t in G corresponds to the case of a system containing an energy source of finite capacity (say a battery or the sun). Surprisingly, in the linear case the finite energy concept is trivial (because an energy source with finite capacity must be nonlinear).

Problem: Show that every linear finite energy operator is S-passive.

Problem: Give an example of a (necessarily nonlinear) finite energy operator which is not S-passive.

The generalization of our various energy concepts to nonlinear operators can be carried out via either or two approaches. First, we can use the same definitions as in the linear case for which we retain the terminology or we can adopt "Lipschitz like" definitions wherein we deal with differences of vectors rather than vectors themselves (SN-1, PR-4). As such, Lipschitz S-passivity is defined by the constraint

$$||E^t(x-y)|| - ||E^tT(x) - E^tT(y)|| \geq 0$$

for all x, y in H and t in G, and similarly for the other passivity and losslessness concepts. From the physical point of view of energy constraints we desire to work with the classical definitions but the Lipschitz definitions are more amenable to mathematical manipulation and are preferable in certain applications (characterizing sensitivity reducing systems as an example).

Problem: Show that a (possibly) nonlinear operator is Lipschitz S-passive if and only if it is same-input same-output causal and has Lipschitz norm less than or equal to one.

A similar result holds for Lipschitz S-lossless operators, but this can be obtained trivially from the linear vase via

Problem: Show that every Lipschitz S-lossless operator is Lipschitz isometric.

Problem: Show that every Lipschitz isometric operator satisfying s0=0 is linear.

In the case of Lipschitz I-passivity we need an additional assumption if a result analogous to that for Lipschitz S-passivity is to be obtained. For this purpose, given any x and y in H we let

$$\alpha_t(x,y) = ||x|| \mathrm{Re}\langle E_t x/||x||, T(x+y)-T(y)\rangle$$

and then may prove:

Problem: Let

$$\lim_{t \to \infty} \frac{\alpha_t(x,y)}{||x||} = 0$$

for all x and y and then show that T is Lipschitz I-passive if and only if it is monotone and same-input same-output causal.

Unlike the Lipschitz passivity conditions the usual passivity conditions tend to degenerate in the nonlinear case even though their physical interpretation may be more viable than that of the Lipschitz passivity concepts (WI-1). The difficulty is due primarily to the fact that the different causality concepts which coincide in the linear and weakly additive cases do not coincide in the general nonlinear case.

Problem: Show that every same-input same-output causal operator with norm less than or equal to one is S-passive.

Problem: Show that every norm causal operator with norm less than or equal to one is S-passive.

Problem: Show that every S-passive operator is zero-input zero-output causal and has norm less than or equal to one.

Problem: Give examples to show that none of the above results are necessary and sufficient for general nonlinear operators.

Problem: Show that a weakly additive operator is S-passive if and only if it is causal (in any sense) and has norm less than or equal to one.

We note that although the concept of S-passivity differs from that of stability only by a scale factor, the above difficulty was not encountered in our nonlinear stability theory since by assuming well-posedness we made an a-priori assumption that the operator was same-input same-output causal (and hence causal in all of the other senses) in that case.

Surprisingly, even though we do not have a necessary and sufficient condition for S-passivity in terms of causality and a norm condition, it is possible to obtain such a result for S-losslessness which unlike Lipschitz S-losslessness is non-trivial in the non-linear case.

Problem: Give an example of a non-linear S-lossless operator.

Problem: Show that an operator is S-lossless if and only if it is isometric and norm causal.

4. Sensitivity of Stability

An interesting problem of system theory is the sensitivity of stability behavior. That is, if a given system is stable, is a small perturbation of that system also stable (DE-3). This is of particular importance in our sensitivity configuration where we may know that our given nominal system is stable but we do not know a-priori if the actual system is stable.

Problem: Let a feedback system with nominal open loop gain K' be well-posed and stable. Then show that there exists an $\varepsilon > 0$ such that if

$$||K' - K|| < \varepsilon$$

the actual system with open loop gain K is also stable.

Consistent with the above, the stable open loop gains form an open set in the uniform operator topology as do the stable plants if one makes the assumption that the feedback term does not change.

Another sensitivity of stability behavior theorem allows arbitrarily large perturbations of K' but assumes them to be of a special form (DE-3).

Problem: Show that if a system with open loop gain K' is well-posed and stable then a system with open loop gain

$$K = K' + C$$

where C is strictly causal is also well-posed and stable.

5. Optimal Controllers

The optimal control problems are, in general, not solvable because of the causality constraint. If, however, this constraint is lifted, one can give a general solution (PR-3).

Problem: Show that for any operator K and d in H the functional

$$J = ||y||^2 + ||e||^2$$

is minimized under the constraint that

$$y = Ke + d$$

by the vector

$$e = -(I+K^*K)^{-1}K^*d$$

There are a number of alternative derivations of the above result using varying amounts of power. One can simply take the Freschet derivative of the functional and set it to zero as per standard optimization theoretic techniques (LA-1). Alternatively, upon observing that J is just the norm of the pair (y,e) in the space H^2 one can apply the projection theorem in this space to obtain the element in the linear variety defined by the constraint equation which minimizes J (PR-3). Finally, a completely elementary proof is possible wherein no "classical" optimization theory whatsoever is required. Here, one uses standard algebraic manipulation and the positive definiteness of (I+K*K) to verify that for any e

$$J(d) \geq \langle d,d\rangle - \langle d,K(I+K^*K)^{-1}K^*d\rangle$$

with equality achieved only for the required value of e.

Since the operator $(I+K^*K)^{-1}K^*$ is not, in general, causal the preceeding result does not yield a solution to the optimal control problem. It, however, indicates the value of the optimal solution and thus we may solve the optimal control problem, under various assumptions on K and D, by finding a causal operator M which is defined on all of H and whose restriction to D coincides with $(I+K^*K)^{-1}K^*$. In one special case, however, such an approach is not needed.

Problem: For a memoryless K and $D \subset H$ show that

$$M = (I+K^*K)^{-1}K^*$$

exists and is a solution to the open loop control problem.

Problem: For a memoryless K and $D \subset H$ show that

$$F = (I+2K^*K)^{-1}K^*$$

exists and is a solution to the feedback control problem.

In general, however, a solution to the optimal control problems requires that we restrict D and simply define M so that its restriction to D coincides with the above M. The class of disturbance spaces, D, for which this is most readily achieved are the predictive subspaces. That is, a subspace D of H such that if

$$E^t x = E^t y$$

for any t and x, y in D then x = y. In essence, the elements of a predictive subspace are completely determined by any truncate. A predictive subspace does not define a sub-resolution space since E does not map D into D, but this property is not needed for the optimal control problems.

Problem: Show that a predictive subspace, D, defines a sub-resolution space if and only if D = 0.

Although on first impression the predictive subspaces may seem to be rather trivial they, in fact, include the classes of disturbance most commonly studied in optimal control problems, and we believe that an assumption of predictability is essential to most classical optimal control theory though such an assumption is never explicitly made.

Problem: Consider the resolution space $L_2(G,K,m^+)$ where K is a Hilbert space with its usual resolution of the identity, and let S(t) be a semi-group of operators (C.C.1) defineed on K for t in G^+ such that

$$\operatorname{limit}_{t \to \infty} ||S(t)|| = 0$$

then show that the subspace D of $L_2(G,K,m^+)$ defined by

$$D = \{f(t);\ f(t) = S(t)x,\ x \text{ in } K\}$$

is predictive.

The subspace of the above problem corresponds to the class of disturbances encountered in the regulator problem wherein the disturbance is the natural response of a differential operator K.

Problem: Show that any subspace of analytic functions (i.e. the restriction of analytic functions to the reals) is a predictive subspace of $L_2(R,R,m)$ with its usual resolution structure.

For disturbances in a predictive subspace we would like to define an optimal controller by the formula

$$M = (I+K^*K)^{-1}K^*P_D$$

where P_D is a projection onto the predictive subspace D. Clearly, this is optimal on D (since it coincides with the general solution for d in D) and it may be rendered causal by choosing the correct projection. Unfortunately, the orthogonal projection is not the correct choice, the desired one being unbounded.

<u>Problem</u>: For any x in H define P_D by

$$P_D x = \begin{cases} d; & \text{if there exists a } d \text{ in } D \text{ and } t \text{ in } G;\ E^t x = E^t d \\ 0; & \text{otherwise.} \end{cases}$$

and show that P_D is well defined for any predictive subspace D. (i.e. show that there is at most one d for which $E^t x = E^t d$).

<u>Problem</u>: Show that the P_D of the above problem is an unbounded linear operator on H which is the identity on D.

<u>Problem</u>: Show that for any linear operator T and any predictive subspace that TP_D is causal.

<u>Problem</u>: For any causal K on (H,E) and predictive subspace D show that

$$M = (I+K^*K)^{-1}K^*P_D$$

is an (unbounded) solution to the open loop optimal control problem. Unfortunately, since M is unbounded the feedback control defined by

$$F = M(I+KM)^{-1}$$

in our equivalence theorem for the open loop and feedback control problems is not assured to exist and will yield an unstable control system when defined. As such, the above result, which except for the unboundedness of M solves the open loop control problem for predictive D does not yield a solution to the feedback control problem.

3. DYNAMICAL SYSTEMS

Unlike the feedback systems of the preceeding chapter wherein each component was represented by a single operator which describes its input-output relation independently of its internal structure, in a dynamical system one deals explicitly with the internal memory structure of the components. Here the evolution of the signals stored in the memory is termed the system dynamics.

There are two alternative approaches to the study of dynamical systems. First, we may assume a-priori a particular representation for the internal memory, studying the system in terms of the properties of this representation as done by Balakrishnan (BL-1, BL-2, BL-3, BL-4, BL-5), Kalman (KM-1, KM-2, KM-3), Mesarovic (ME-1), and Zadeh (ZD-1, ZD-2). The alternative approach which we adopt is to begin with an abstract (input-output) operator on a resolution space and axiomatically construct the feasible internal memory structures. In fact, under appropriate assumptions this internal structure is essentially unique and hence the internal dynamics of the system is, indeed, an operator invariant. The dynamical system concept presented is essentially that of DeSantis and Porter (DE-1, DE-8) and the author (SA-5) and is, like the causality and stability concepts already presented, inherently resolution space in nature with the concepts undefined (undefinable) in a classical operator theoretic context.

We begin with the formulation of the axiomatics required to construct a state decomposition of a given operator on a resolution space. Given such a decomposition the traditional dynamical system concepts such as transition operators and state evolution formulae are developed, the latter of these taking the form of the Cauchy integral which has already dominated so much of our theory. We then consider the formulation of such classical concepts as controllability, observability and (internal) stability in the resolution space setting, and finally the regulator problem is introduced.

A. State Decomposition

1. Definitions and Examples

Conceptually the state space of an operator is the "memory" internal to the system the operator represents, and one assumes that an input prior to time t is mapped by the system into an element of the state space (which is stored in the memory) which is the sole manifestation of the part of the input prior to time t on the output after time t. Of course, there are many internal memory structures which have the same effect on the system inputs and outputs; hence, if one simply represents a system by an operator on a resoluiton space without an a-priori model of the memory in the actual system, the state space is non-unique. Rather, we give an axiomatic characterization of the admissible state-space structures for the system without fixing a particular structure. This is achieved via the state decomposition of an operator T which is a triple (S,G(t),H(t)) where S is a Hilbert space, the state-space, and G(t) and H(t) are bounded oeprator valued (Haar measurable) functions defined on G such that

i) G(t) maps H into S and satisfies the equality

$$G(t) = G(t)E^{t}$$

for all t in G.

ii) H(t) maps S into H and satisfies

$$H(t) = E_{t}H(t)$$

for all t in G.

iii)

$$H(t)G(t) = E_{t}TE^{t}$$

for all t in G. i.e., the diagram

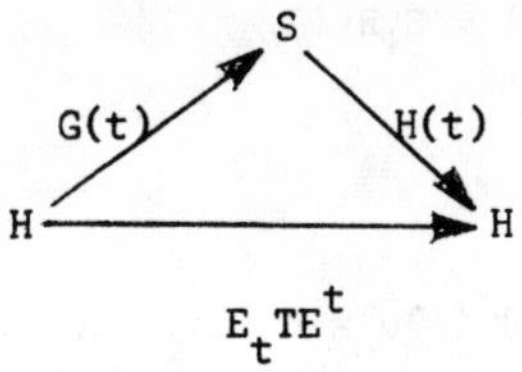

commutes for all t in G.

Here (H,E) is an arbitrary resolution space which will remain fixed throughout the development.

Note that every operator, T, on (H,E) admits the state decomposition (H,E^t,E_tT).

Consider the operator on $L_2(Z,R^n,m^+)$ with its usual resolution structure defined by the difference equation

$$X_{k+1} = AX_k + Bu_k$$

$$y_k = CX_k$$

with $X_o = 0$, X_k an m-vector and A, B, C matrices of appropriate dimension. Now for any i in Z we let G(i) map $L_2(Z,R^n,m^+)$ into $S = R^m$ via

$$G(i)u = \sum_{j=0}^{i-1} A^j Bu_{i-j} = X_i$$

and H(i) maps $S = R^m$ to $L_2(Z,R^n,m^+)$ via

$$H(i)X = (0,0,\ldots,0,CX,CAX,CA^2X,\ldots)$$

where H(i)X is assured to be in $L_2(Z,R^n,m^+)$ if the given difference operator is well defined in this space. Now by construction this satisfies

$$G(i) = G(i)E^i$$

and

$$H(i) = E_i H(i)$$

whereas

$$H(i)G(i)u = H(i)X_i = (,\ldots,0,CX_o,CAX_i,\ldots)$$

which coincides with $E_i TE^i$ for the given difference operator T. As such, the classical difference equations state model does, indeed, define an abstract state decomposition (ZD-3).

Similarly, for the differential operator

$$\dot{X} = ZX + Bu$$

$$y = CX$$

on $L_2(R,R^n,m^+)$ with its usual resolution structure and X in R^m we may define a state decomposition with $S = R^m$ via

$$G(t)u = \int_o^t e^{A(t-q)}Bu(q)dq = X(t)$$

and

$$(H(t)X)(s) = Ce^{A(s-t)}XU(s-t)$$

hence the classical differential equation state concept also defines an abstract state decomposition (ZD-3).

2. A Uniqueness Theorem

Although the state decomposition is in general highly non-unique we are primarily interested in the case of a minimal state decomposition wherein H(t) is one-to-one and G(t) is onto for all t in G which is essentially unique when it exists (DE-1, SA-5). In this case, following the differential operator notation we say that an operator on a resolution space (H,E) is <u>regular</u> if it admits a minimal state decomposition (ZD-3, SA-10).

<u>Theorem</u>: Let $(S,G(t),H(t))$ and $(\underline{S},\underline{G}(t),\underline{H}(t))$ be two minimal state decompositions of an operator T on (H,E). Then there exists a unique family of (algebraic) isomorphisms

$$K(t):S \rightarrow \underline{S}$$

defined for all t in G such that the diagram

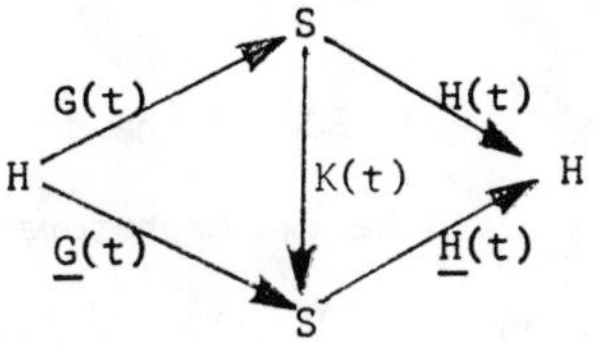

commutes for all t in t.

Proof: Since G(t) is onto for any x in S there exists a y in H such that $x = G(t)y$ and we let K(t) be defined by the equality

$$K(t)x = \underline{G}(t)y$$

Now if y' is another element of H such that $G(t)y' = x$ then since

$$H(t)G(t) = \underline{H}(t)\underline{G}(t) = E_tTE^t$$

we have

$$\underline{H}(t)\underline{G}(t)y'=H(t)G(t)y'=H(t)x=H(t)G(t)y=\underline{H}(t)\underline{G}(t)y$$

and since $\underline{H}(t)$ is one-to-one this implies that

$$\underline{G}(t)y' = \underline{G}(t)y$$

hence verifying that the operator K(t) is independent of the choice of y in the preimage of G(t) and thus well defined.

Now for any y in H with $x = G(t)y$

$$K(t)G(t)y = K(t)x = \underline{G}(t)y$$

since y is in the preimage of x under G(t). As such the left hand side of the diagram commutes. On the other hand, for any x in S with y in the preimage of x under G(t)

$$\underline{H}(t)K(t)x=\underline{H}(t)\underline{G}(t)y=H(t)G(t)y=H(t)x$$

verifying that the right hand side of the diagram also commutes.

Now let $K_1(t)$ and $K_2(t)$ be two operators which render the diagram commutative. Then for any x in S

$$\underline{H}(t)K_1(t)x=H(t)x=\underline{H}(t)K_2(t)x$$

and since $\underline{H}(t)$ is one-to-one this implies that $K_1(t)x = K_2(t)x$ verifying the uniqueness of K(t).

Finally, for any $\underline{x}$ in $\underline{S}$ since $\underline{G}(t)$ is onto there exists y in H such that $\underline{x} = \underline{G}(t)y$ and

$$\underline{H}(t)\underline{x}=\underline{H}(t)\underline{G}(t)y=H(t)G(t)Y=H(t)x$$

where $x = G(t)y$. Now by the commutativity of the left side of the diagram we have

$$\underline{H}(t)K(t)x=H(t)x=H(t)G(t)y=\underline{H}(t)\underline{G}(t)y=\underline{H}(t)\underline{x}$$

and since $\underline{H}(t)$ is one-to-one this implies that $K(t)x = \underline{x}$ verifying that $\underline{x}$ is in the range of K(t) and hence that it is onto. Similarly, if

$$K(t)x = K(t)x'$$

let y and y' be in the preimage of x and x', respectively, under G(t); hence, by the construction of K(t) we have

$$\underline{G}(t)y=K(t)x=K(t)x'=\underline{G}(t)y'$$

and thus

$$H(t)x'=H(t)G(t)y'=\underline{H}(t)\underline{G}(t)y'=\underline{H}(t)\underline{G}(t)y$$

$$=H(t)G(t)y=H(t)x$$

Now since H(t) is one-to-one x = x' verifying that K(t) is also one-to-one and thus completing the proof of the theorem.

Note that the theorem in essence implies that a minimal state decomposition is unique except for the "change of variable" defined by K(t) and hence our desired "essential uniqueness" condition has been verified.

Since the minimal state decomposition, when it exists, is uniquely determined by the operator T (i.e., it is an invariant of T) one would expect to be able to characterize the properties of T via its minimal state decomposition. This is, indeed, the case as is illustrated by the following theorem.

<u>Theorem</u>: A regular operator, T, on (H,E) has a zero minimal state decomposition if and only if it is anti-causal.

<u>Proof</u>: If T is anti-causal

$$E_tT = E_tTE_t$$

hence

$$E_tTE^t = E_tTE_tE^t = 0$$

whence (0,0,0) is indeed a state decomposition. Conversely, if T has a zero minimal state decomposition then

$$E_t T E^t = 0'0 = 0$$

verifying that T is anti-causal and completing the proof.

Note that as a corollary to the above one may state that a causal operator has a zero minimal state decomposition if and only if it is memoryless (SA-5, DE-1, DE-8).

3. Transition Operators

In the classical theory of differential and difference equations a significant role is played by the transition operators $e^{A(t-q)}$ and $A^{(i-j)}$, respectively (ZD-3). Analogous operators, indeed, exist in our general theory as per the following theorem (SA-5).

Theorem: Let (S,G(t),G(t)) be a minimal state decomposition of a regular operator, T, on (H,E). Then for any $q \leq t$ in G there exists a unique family of homomorphisms $\emptyset(t,q)$ such that the diagram

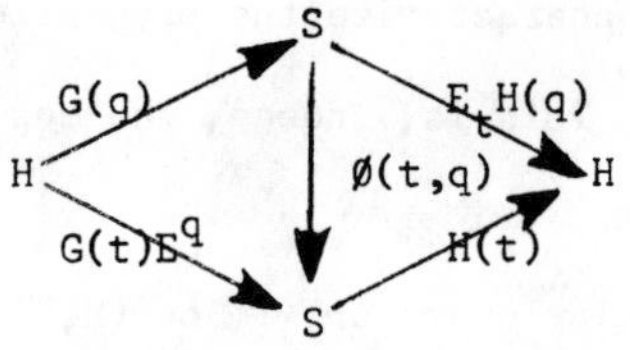

commutes.

Proof: First, since $q \leq t$

$$E_t H(q)G(q) = E_t E_q T E^q = E_t T E^q = E_t T E^t E^q = H(t)G(t)E^q$$

verifying that the outer diamond is commutative and hence that the statement of the problem is well-defined. Now for any x in S since G(q) is onto there exists y in H such that x = G(q)y and we may define $\emptyset(t,q)$ by

$$\emptyset(t,q)x = G(t)E^q y$$

which is well defined for if we let y' be another element of H such that $x = G(q)y'$ then since the outer diamond commutes

$$H(t)G(t)E^q y' = E_t H(q)G(q)y' = E_t H(q)x = E_t H(q)G(q)y$$

$$= H(t)G(t)E^q y$$

and since H(t) is one-to-one this implies that

$$G(t)E^q y' = G(t)E^q y$$

and hence $\emptyset(t,q)$ is well defined.

Now for any y in H with $x = G(q)y$

$$\emptyset(t,q)G(q)y = \emptyset(t,q)x = G(t)E^q y$$

verifying the commutativity of the left hand triangle. Similarly, for any x in S with y in the preimage of x under G(q)

$$H(t)\emptyset(t,q)x = H(t)G(t)E^q y = E_t H(q)G(t)y = E_t H(q)x$$

verifying the commutativity of the right hand triangle.

Finally, if $\emptyset_1(t,q)$ and $\emptyset_2(t,q)$ are two mappings rendering the diagram commutative then for any x in S the commutativity of the right hand triangle yields

$$H(t)\emptyset_1(t,q)x = E_t H(q)x = H(t)\emptyset_2(t,q)x$$

and the fact that H(t) is one-to-one implies that

$$\emptyset_1(t,q)x = \emptyset_2(t,q)x$$

verifying the uniqueness of $\phi(t,q)$ and thus completing the proof of the theorem.

A family of operators $\phi(t,q)$, satisfying the commutativity condition of the preceeding theorem for $t \geq q$ and equal to zero for $t < q$ are termed the transition operators for the state decomposition $(S,G(t),H(t))$. Of course, the theorem implies that a unique family of transition operators exists for a minimal state decomposition though, in fact, transition operators may exist for non-minimal decompositions. For instance,

$$\phi(t,q) = \begin{cases} I & t \geq q \\ 0 & t < q \end{cases}$$

is always a valid set of transition operators for the decomposition (H,E^t,E_tT) of an operator T on (H,E).

For the difference operator

$$X_{k+1} = AX_k + Bu_k$$

$$y_k = CX_k$$

with the state decomposition of (3.A.1) a set of transition operators are given by

$$\phi(i,j) = A^{i-j}$$

whereas for the differential operator

$$\dot{X} = AX + Bu$$

$$y = CX$$

with the state decomposition of (3.A.1) the transition operators are, as in classical differential equation theory, given by

$$\emptyset(t,q) = e^{A(t-q)}$$

Although we have defined $\emptyset(t,q)$ for a fixed t and q independently of any other time the transition operators for different times are, in fact, related (SA-5, DE-1, DE-8).

Theorem: Let $\emptyset(t,q)$ be the transition operators associated with a minimal state decomposition of a regular operator on (H,E). Then

$$\emptyset(t,t) = I$$

for all t in G and

$$\emptyset(t,q)\emptyset(q,r) \geq \emptyset(t,r)$$

whenever $r \leq q \leq t$ are in G.

Proof: Since $G(t) = G(t)E^t$ and $H(t) = E_t H(t)$ whenever t = q the diagram defining $\emptyset(t,t)$ reduces to

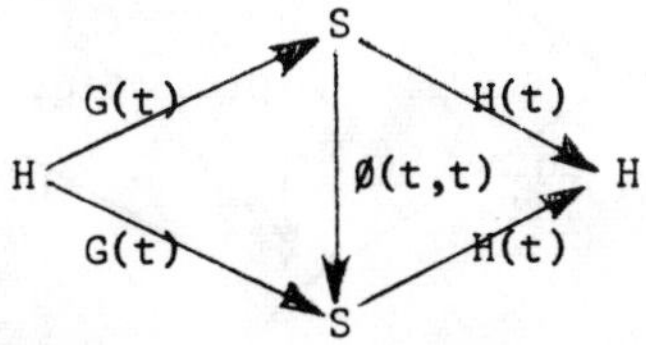

Now this is clearly rendered commutative by $\emptyset(t,t) = I$ and since $\emptyset(t,t)$ is unique we have verified the required equality.

Now given $r \leq q \leq t$ consider the diagram

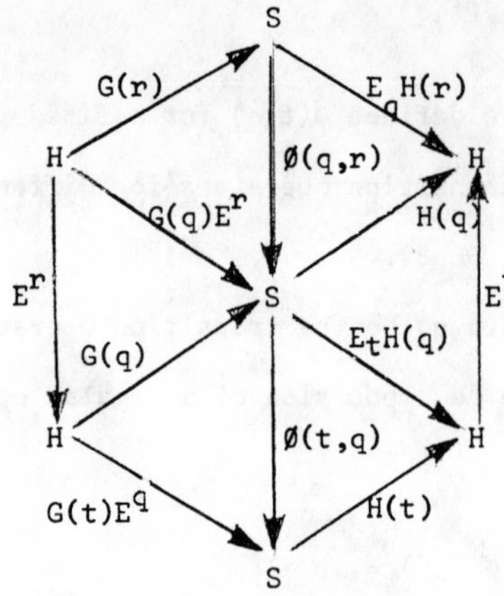

where the upper and lower diamonds are commutative via the definition of $\emptyset(t,q)$ and $\emptyset(q,r)$ and the left and right hand triangles are trivially commutative. As such, the entire diagram is commutative and if we consider only the outer rim and the center line it reduces to the commutative triangle

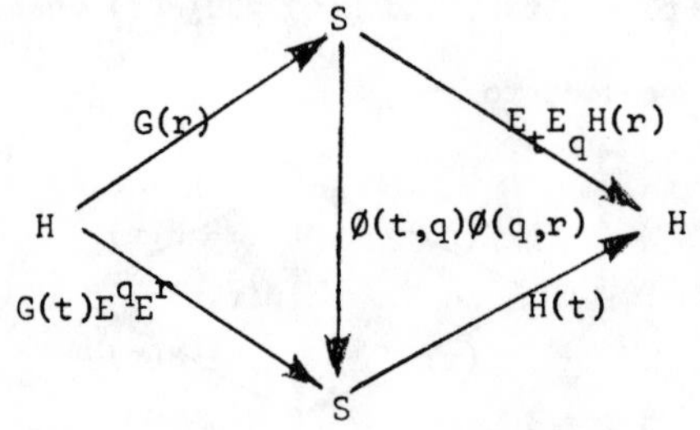

Finally, since $r \leq q \leq t$ we have $E^q E^r = E^r$ and $E_t E_q = E_t$; hence, the above is precisely the diagram defining $\emptyset(t,r)$ which being the unique operator which renders the diagram commutative must equal $\emptyset(t,q)\emptyset(q,r)$ since this also renders the diagram commutative. The proof of the theorem is therefore complete.

4. State Trajectories

Given a system with state decomposition $(S,G(t),H(t))$, for any input u the state induced at t is $G(t)u$, hence we may define the state trajectory to be the Borel measurable function (since G(t) is measurable)

$$x(t) = G(t)u$$

Since $G(t) = G(t)E^t$, x(t) is clearly determined entirely by $E^t u$. In fact, however, if our intuitive interpretation of the state concept to the effect that "the state at time s contains all relevant information concerning the effect of inputs applied before s on outputs after s" is correct we should be able to compute x(t) in terms of x(s) and E(q,t)u if $t \geq s$. Indeed, this is the case as per the following theorem.

<u>Theorem</u>: Let T be a linear operator on (H,E) with state decomposition (S,G(t),H(t)) and transition operators $\emptyset(t,q)$ then for any input u and $t \geq s$ in G

$$x(t) = SRC\int_{(s,t)} \emptyset(t,q)G(q)dE(q)u + \emptyset(t,s)x(s)$$

Proof : Let

$$s = t_o < t_1 < t_2 < \dots < t_n = t$$

be a partition of (s,t) and for any u in H let

$$u = \overline{u} + \underline{u} = \overline{u} + \sum_{i=0}^{n} u_i$$

where $\overline{u} = E_t u$, $\underline{u} = E^t u$ and $u_i = [E^{t_i} - E^{t_{i-1}}]u$ for i=1,2,...,n and $u_o = E^s u = E^{t_0} u$. Now since $G(t) = G(t)E^t$ we have

$$x(t)=G(t)u=G(t)E^t u=G(t)\underline{u}=G(t)\sum_{i=0}^{n} u_i$$

$$=\sum_{i=1}^{n} G(t)[E^{t_i}-E^{t_{i-1}}]u+G(t)E^s u$$

$$=\sum_{i=1}^{n} G(t)E^{t_i}[E^{t_i}-E^{t_{i-1}}]u+G(t)E^s u$$

Now upon substitution of the equality $G(t)E^q = \emptyset(t,q)G(q)$ which follows from the defining diagram for $\emptyset(t,q)$ this becomes

$$x(t)= \sum_{i=1}^{n} \emptyset(t,t_i)G(t_i)[E^{t_i}-E^{t_{i-1}}]u+\emptyset(t,s)x(s)$$

and since the value of the partial sum is independent of the partition upon taking limits over all partitions we obtain

$$x(t)=SRC\int_{(s,t)} \emptyset(t,q)G(q)dE(q)u+\emptyset(t,s)x(s)$$

as required.

A natural variation on the preceeding theorem is to take the limit as s goes to $-\infty$ whence if the limit of the initial condition term goes to zero we obtain a representation of x(t) directly in terms of u. In fact, however, an inspection of the proof of the preceeding theorem will reveal that such a representation may be obtained directly if one takes s to be $-\infty$ in which case the u_o term drops out. We then have

Theorem: Let T be a linear operator on (H,E) with state decomposition (S,G(t),H(t)) and transition operators $\emptyset(t,q)$. Then for any input u and t in G

$$x(t) = SRC\int_{(-\infty,t)} \emptyset(t,q)G(q)dE(q)u$$

5. An Invariance Theorem

Modulo our equivalence concept we already know that a minimal state decomposition is uniquely determined by an operator, hence it is an operator invariant. It remains, however, to determine whether or not it is a complete set of invariants for the operator; i.e., can the operator be reconstructed from the state decomposition. In general, the answer is no since the minimal state decomposition for an anti-causal operator is zero and hence contains no information about the operator. In fact, we have (SA-5, DE-1):

<u>Theorem</u>: Let a regular operator, T, have a well defined strongly strictly causal part (1.F.2)

$$C = SLC\int_G dE(t)TE^t$$

Then any minimal state decomposition (S,G(t),H(t)) for T is a complete set of invariants for C (i.e., the state decomposition is determined entirely by C and C can be reconstructed from the state decomposition.)

Proof: If C exists then

$$A=T-C=T-SLC\int_G dE(t)TE^t=SLC\int_G dE(t)T-SLC\int_G dE(t)TE^t$$

$$=SLC\int_G dE(t)TE_t$$

is anti-causal. Hence

$$E_t TE^t=E_t CE^t+E_t AE^t=E_t CE^t$$

showing that (S,G(t),H(t)) which is defined via $E_t TE^t$ is in fact dependent only on C.

Conversely, given a minimal state decomposition of T, (S,G(t),H(t)) with a transition operator $\emptyset(t,q)$ (which are assured to exist via minimality) we have

$$C=SLC\int_G dE(t)TE^t=SLC\int_G dE(t)E_t TE^t$$

Now

$$E^t = SRC\int_{(-\infty,t)} dE(r) = SRC\int_{(-\infty,t)} E^r dE(r)$$

hence

$$C = SLC\int_G dE(t)E_t TE^t = SLC\int_G dE(t)E_t T[SRC\int_{(-\infty,t)} E^r DE(r)]$$

$$= SLC\int_G dE(t)[SRC\int_{(-\infty,t)} E_t TE^r dE(r)]$$

Since this last integral is only over the interval $(-\infty,t)$ for any r in this range of integration, $E_t = E_t E_r$; hence we have

$$C = SLC\int_G dE(t)[SRC\int_{(-\infty,t)} E_t E_r TE^r dE(r)]$$

$$= SLC\int_G dE(t)[SRC\int_{(-\infty,t)} E_t H(r)G(r)dE(r)]$$

Finally, since

$E_t H(r) = H(t)\emptyset(t,r)$ this becomes

$$C = SLC\int_G dE(t)[SRC\int_{(-\infty,t)} H(t)\emptyset(t,r)G(r)dE(r)]$$

$$= SLC\int_G dE(t)H(t)[SRC\int_{(-\infty,t)} \emptyset(t,r)G(r)dE(r)]$$

which is a representation of C entirely in terms of the state decomposition (and the transition operators which are derived from the state decomposition) as required to complete the proof of the theorem.

A comparison of our two preceding theorems yields the following.

Theorem: Let T be a regular strongly strictly causal operator with minimal state decomposition $(S,G(t),H(t))$ and let $x(t)$ be the state trajectory resulting from an input u. Then

$$Tu = SLC\int_G dE(t)H(t)x(t)$$

In essence, the theorem says that we can factor a strongly strictly causal operator through its state space via

$$T = JP$$

where P is the operator mapping H into a space of S valued (state trajectory) functions via

$$(Pu)(t) = x(t) = SRC\int_{(-\infty,t)} \phi(t,q)G(q)dE(q)u$$

and J is the operator mapping the space of S valued functions into H via

$$Jx = Tu = SLC\int_{G} dE(t)H(t)x(t)$$

Note that unless we make some sort of stability assumption we have no assurance that the state trajectories lie in a Hilbert space; hence, J and P are not formally operators.

B. Controllability, Observability, and Stability

1. Controllability

Conceptually, the state space is a model of the "internal" memory of an operator (which is essentially unique if one assumes minimality) and hence it is natural to attempt to determine the relationship between the internal and external properties of a given operator. The relationship between operator inputs and the state space manifests itself in the concept of controllability wherein we say that a state decomposition $(S,G(t),H(t))$ is controllable if there exists a positive constant M such that for each x in S and t in G there exists a u in H with $||u|| \leq M||x||$ such that

$$x = G(t)u$$

In essence, controllability implies that $G(t)$ is onto and also possesses a bounded right inverse. A somewhat stronger concept is a uniformly controllable state decomposition wherein we require that there exist a positive constant M and a $\delta > 0$ in G such that for each x in S and t in G there exists a u in H with $||u|| \leq M||x||$ such that

$$x = G(t)E_{t-\delta}u$$

i.e., $G(t)E_{t-\delta}$ has a (uniformly in t) bounded right inverse.

The controllable state decompositions are characterized by the following theorem.

Theorem: A state decomposition $(S,G(t),H(t))$ is controllable if and only if the operators $G(t)G(t)^*$ are uniformly positive definite. (i.e., there exists an $\varepsilon > 0$ such that $\langle x,G(t)G(t)^*x\rangle \geq \varepsilon \langle x,y\rangle$ for all x and t.)

Proof: If (S,G(t),H(t)) is controllable the G(t) all have right inverses with norms uniformly bounded by M. Thus since

$$I=I^*=[G(t)G(t)^{-R}]^*=G(t)^{-R^*}G(t)^*$$

the G(t)* all have left inverses

$$G(t)^{*-L}=G(t)^{-R^*}$$

with norms also uniformly bounded by M. We thus have

$$<x,G(t)G(t)^*x>=||G(t)^*x||^2=\frac{1}{||G(t)^{*-L}||^2}||G(t)^{*-L}||^2||G(t)^*x||^2$$

$$\geq\frac{1}{M^2}||G(t)^{*-L}G(t)^*x||^2=\varepsilon<x,x>$$

where $\varepsilon = 1/M^2$

Conversely, if G(t)G(t)* is uniformly positive definite, then

$$\frac{1}{M^2}||x||^2=\varepsilon<x,x>\leq<x,G(t)G(t)^*x>=||G(t)^*x||^2$$

which implies that the G(t)* have left inverses with norms uniformly bounded by M and hence the G(t) have right inverses with norms uniformly bounded by M and as such are controllable.

An argument similar to that for the controllability theorems yields the following characterization for the uniformly controllable state decompositions.

Theorem: A state decomposition, (S,G(t),H(t)) is uniformly controllable if and only if there exists a $\delta > 0$ in G such that the operators $G(t)E_{t-\delta}G(t)^*$ are uniformly positive definite.

2. Observability

The natural dual of the controllability concept is that of observability which characterizes the coupling between the state space of an operator and its output. We say that a state decomposition $(S,G(t),H(t))$ is <u>observable</u> if there exists a positive constant M such that for all x and t

$$||H(t)x|| \geq M||x||$$

and it is <u>uniformly observable</u> if there exists a positive constant M and a $\delta > 0$ in G such that

$$||E^{t+\delta}H(t)x|| \geq M||x||$$

In essence, the observability concepts require that the $H(t)$ or $E^{t+\delta}H(t)$ are appropriately left invertible.

The characterization theorems for observability and uniform observability dual to those for controllability and uniform controllability are as follows.

<u>Theorem</u>: A state decomposition $(S,G(t),H(t))$ is observable if and only if the operators $H(t)^*H(t)$ are uniformly positive definite.

<u>Theorem</u>: A state decomposition $(S,G(t),H(t))$ is uniformly observable if and only if there exists a $\delta > 0$ such that the operators $H(t)^*E^{t+\delta}H(t)$ are uniformly positive definite.

3. <u>Examples</u>

For the differential operator

$$\dot{X} = A(t)X + B(t)u$$

$$y = C(t)X$$

with $X(0) = 0$ defined on $L_2(R,R^n,m^+)$ with its usual resolution structure a state decomposition with $S = R^m$ (m the dimension of X) is given by

$$G(t)u = X(t) = \int_0^t \phi(t,q)B(q)u(q)dq$$

and

$$[H(t)X(t)](s) = C(s)\phi(s,t)X(t)$$

where $\phi(t,q)$ is the usual family of transition operators (ZD-3) for the time-varying differential equations. As such, straightforward algebraic manipulations will reveal that $G(t)G(t)^*$ is uniformly positive definite if and only if the family of m by m matrices

$$\int_0^t \phi(t,q)B(q)B(q)'\phi(t,q)'dq$$

is uniformly positive definite and, similarly $H(t)^*H(t)$ is uniformly positive definite if and only if the family of matrices

$$\int_t^\infty \phi(q,t)'C(q)'C(q)\phi(q,t)dq$$

is uniformly positive definite (ZD-3). As such, even though we are dealing with infinite dimensional problems, our test for controllability and observability is only finite dimensional for systems with a finite dimensional state space.

4. Internal Stability

The controllability and observability concepts, in essence characterize the algebraic coupling between the system inputs and outputs and its state space. The continuity of these relationships is characterized by their (internal) stability. For a state decomposition $(S,G(t),H(t))$ we say that it is bounded-input bounded-state (BIBS) stable if there exists a positive constant M such that

$$||G(t)|| \leq M$$

for all t in G and we say that it is <u>bounded-state bounded-output (BSBO) stable</u> if there exists a positive constant M such that

$$||H(t)|| \leq M$$

for all t in G.

Three additional modes of stability characterize the continuity of the transition operator, $\emptyset(t,q)$, associated with $(S,G(t),H(t))$ when they exist. For this purpose we say that the state decomposition is <u>Lyapunov (L) stable</u> if there exists a positive constant M such that

$$||\emptyset(t,q)|| < M$$

for all t and q in G. We say that it is <u>asymptotically (A) stable</u> if

$$\lim_{(t-q)\to\infty} || \emptyset(t,q)|| = 0$$

and we say that it is <u>exponentially (E) stable</u> if there exist a strictly positive real character $r > 0$ and a positive constant M such that

$$||\emptyset(t,q)|| \leq M(r,t-q)_p$$

So as to distinguish the above stability concepts from the (external) stability studied in the context of feedback systems we term the above internal stability concepts though if there is no possibility of ambiguity we simply say stability. Also, note that although the minimal state decomposition of an operator is unique our equivalence concept for state decompositions is algebraic rather than analytic in nature and hence stability is not an operator invariant even for minimal decompositions.

5. Stability Theorems

In general a state decomposition may contain information which is totally unrelated to the operator it decomposes and hence no constraints on the operators by themselves will suffice to assure the internal stability of its state decomposition. If, however, we make certain controllability and/or observability assumptions about the state decomposition which have the effect of insuring a tight coupling between the operator and its state decomposition then the boundedness of the operator itself suffices to imply the internal stability of the state decomposition (SA-5, WL-2). Of course, if we deal only with causal operators, then boundedness is equivalent to (external) stability (2.B.1) and hence with the appropriate controllability and/or observability conditions external stability implies internal stability. Our first theorem of this type is:

Theorem: Let (S,G(t),H(t) be a state decomposition of a bounded operator T. Then (S,G(t),H(t)) is BIBS stable if it is observable.

Proof: If (S,G(t),H(t)) is not BIBS stable, there exists a bounded sequence of inputs u_i and t_i in G such that the sequence of states

$$x_i = G(t_i)u_i$$

is unbounded. Now since $G(t_i) = G(t_i)E^{t_i}$ we may assume that $u_i = E^{t_i}u_i$ whence the norm of the response of T applied to u_i is

$$||Tu_i||^2=||TE^{t_i}u_i||^2=||E^{t_i}TE^{t_i}u_i+E_{t_i}TE^{t_i}u_i||^2$$

$$\geq||E_{t_i}TE^{t_i}u_i||^2=||H(t_i)G(t_i)u_i||^2=||H(t_i)x_i||^2$$

Now by observability

$$||H(t_i)x_i|| \geq M||x_i||$$

hence the sequence of norms $||Tu_i||$ is unbounded since the norms of the x_i are unbounded. Observability thus implies BIBS stability and the proof of the theorem is complete.

An argument dual to that of the preceeding theorem yields

<u>Theorem</u>: Let (S,G(t),H(t)) be a state decomposition of a bounded operator T. Then (S,G(t),H(t)) is BSBO stable if it is controllable.

<u>Theorem</u>: Let (S,G(t),H(t)) be a state decomposition of an operator T with a specified family of transition operators $\emptyset(t,q)$. Then (S,G(t),H(t)) is L stable if it is both controllable and observable.

Proof: If $\emptyset(t,q)$ is not L stable there exists a bounded sequence of states x_i and sequences t_i and q_i in G such that

$$\emptyset(t_i,q_i)x_i$$

is unbounded. Now by controllability there exists a bounded sequence of inputs u_i such that

$$x_i = G(q_i)u_i = G(q_i)E^{q_i}u_i$$

Thus the unbounded sequence of states

$$z_i = \emptyset(t_i,q_i)x_i = \emptyset(t_i,q_i)G(q_i)E^{q_i}u_i = G(T_i)E^{q_i}u_i$$

results from the bounded sequence of inputs $E^{q_i}u_i$ showing that (S,G(t),H(t)) is not BIBS stable and thereby contradicting observability. The proof of the theorem is thus complete.

In general, definitive theorems on A and E stability require some assumption on the properties of G, for instance that it has an archemedian order such as R or Z. In this case arguments similar to those used in the preceeding yield.

<u>Theorem</u>: Let G be an archemedian ordered group and (S,G(t),H(t)) be a state decomposition of an operator T with a specified family of transition operators $\emptyset(t,q)$. Then if (S,G(t),H(t)) is either controllable and uniformly observable or uniformly controllable and observable $\emptyset(t,q)$ is E stable (and hence also A stable).

C. The Regulator Problem

1. Definition

In many practical control problems rather than deriving the control from the system output it is derived from the state of the system which is often accessible from the physical memory elements in the system. For this purpose we let C be a strongly strictly causal operator on a resolution space (H,E) and we let (S,G(t),H(t)) be a state decomposition of C with exponentially stable transition operators $\emptyset(t,q)$. Now for any initial time s we let (H_s,E) be the sub-space of (H,E) composed of vectors satisfying $E^s u = 0$ and we let C_s be the restriction of C to (H_s,E). Similarly, we denote by $(\underline{H}_s,\underline{E})$ the subspace of $L_2(G,S,m)$ composed of functions which are zero for t less than s together with its usual resolution structure. Now via our theory for the factorization of strictly causal operators through their state trajectories (3.A.5), we may factor C_s via

$$C_s = J_s P_s$$

where P_s is the operator mapping (H_s,E) to $(\underline{H}_s,\underline{E})$ defined by

$$x(t) = SRC\int_{(s,t)} \emptyset(t,q)G(q)dE(q)u$$

for any u in H_s and J_s is defined via

$$J_s x = SLC\int_{(s,\infty)} dE(t)H(t)x(t)$$

as a mapping from $(\underline{H}_s,\underline{E})$ to (H_s,E). Here the exponential stability of $\emptyset(t,q)$ assures that P_s is a well defined operator with values in $(\underline{H}_s,\underline{E})$. Finally, for any element, x_s, of the state space, S, we denote by $d_s(x_s)$ the vector in $(\underline{H}_s,\underline{E})$ defined by

$$d_s(x_s)(t) = \emptyset(t,s)x_s$$

Using the above defined objects the state feedback configuration required for the regulator problem is that shown below. Here F_s is a

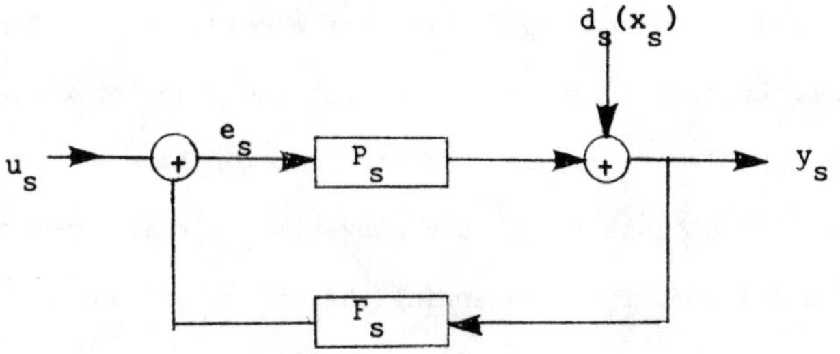

Figure: State Feedback Configuration for the Regulator Problem

causal feedback operator and $d_s(x_s)$ is the disturbance which we desire to regulate. P_s and F_s form a feedback system in the sense previously defined and all of our feedback system theory applies to that part of the above configuration without modification. Moreover, with this particular disturbance, the regulator problem is simply the feedback control problem (2.D.1) for this special configuration and form of disturbance. That is, we desire to find a causal operator F_s such that the feedback system with open loop gain P_sF_s is well-posed and stable and the functional

$$J_s = ||y_s||_s^2 + ||e_s||_s^2$$

is minimized under the constraints

$$y = P_se_s + d_s(x_s)$$

$$e_s = F_sy_s$$

for all x_s in S.

As in our study of the general feedback control problem we are not interested in actually computing the optimal feedback controller but rather in determining its qualitative properties. In particular, we are interested in those properties which are applicable to the regulator problem but not the general feeback control problem. The most important of these is that in the regulator problem F_s may be taken to be memoryless rather than causal. The intuitive basis for this observation is that since the present state contains all information concerning the effect of past inputs on future outputs and P_s is causal there is nothing to be gained by feeding back information from past values of the state (and by causality it is impossible to feedback information from future values of the state). Formal proofs of this intuitively obvious fact are, however, few and far between even for highly specialized cases (KM-4). It is such a proof which is the primary object of the remainder of this section.

2. Principle of Optimality Lemma

As a first step in our proof that memoryless feedback suffices for the solution of the regulator problem we have the following "Principle of Optimality" like lemma. For this purpose given any $r \geq s$ we let (H_r,E), $(\underline{H}_r,\underline{E})$, P_r, etc. be the spaces and operators analagous to (H_s,E), $(\underline{H}_s,\underline{E})$, P_s, etc. but initialized at r rather than s.

Lemma: Let $r \geq s$ and e_s in H_s minimize the functional

$$J_s = ||y_s||_s^2 + ||e_s||_s^2$$

under the constraint

$$y_s = P_s e_s + d_s(x_s)$$

for some x_s in S. Then $e_r = E_r e_s$ minimizes the functional

$$J_r = ||y_r||_r^2 + ||e_r||_r^2$$

under the constraint

$$y_r = P_r e_r + d_r(x_r)$$

where x_r in S is defined by

$$x_r = \int_{(s,r)} \phi(r,q)G(q)dE(q)e_s + \phi(r,s)x_s$$

Proof: Assume that an $e_r' \neq e_r$ in H_r exists such that

$$||y_r'||_r^2 + ||e_r'||_r^2 < ||y_r||_r^2 + ||e_r||_r^2$$

where

$$y_r' = P_r e_r' + d_r(x_r)$$

and consider the vectors

$$e_s' = E^r e_s + e_r'$$

in H_s and

$$y_s' = P_s e_s' + d_s(x_s)$$

Now since P_s is causal and $E^r e_s = E^r e_s'$ we have $\underline{E}^r y_s = \underline{E}^r y_s'$. Also for any $t \geq r$ hence we have

$$y_s'(t) = P_s e_s' + d_s(x_s) = SRC\int_{(s,t)} \phi(t,q)G(q)dE(q)e_s' + \phi(t,s)x_s$$

$$= SRC\int_{(r,t)} \phi(t,q)G(q)dE(q)e_r' + SRC\int_{(s,r)} \phi(t,qG(q)dE(q)e_s$$

$$+ \emptyset(t,s)x_s = \mathrm{SRC}\int_{(r,t)} \emptyset(t,q)G(q)dE(q)e'_r$$

$$+ \emptyset(t,r)\left(\mathrm{SRC}\int_{(s,r)} \emptyset(r,q)G(q)dE(q)e_s + \emptyset(r,s)x_s\right)$$

$$= \mathrm{SRC}\int_{(r,t)} \emptyset(t,q)G(q)dE(q)e'_r + \emptyset(t,r)x_r = y'_r(t)$$

hence $\underline{E}_r y'_s = \underline{E}_r y'_r$. Via a similar argument we also have $\underline{E}_r y_s = \underline{E}_r y_r$. Upon summarizing the above we thus have the following equalities.

$$E^r e_s = E^r e'_s$$

$$\underline{E}^r y_s = \underline{E}^r y'_s$$

$$E_r e'_s = E_r e'_r$$

$$E_r e_r = E_r e_s$$

$$\underline{E}_r y'_s = \underline{E}_r y'_r$$

$$\underline{E}_r y_s = \underline{E}_r y_r$$

and the inequality

$$||y'_r||^2_r + ||e'_r||^2_r < ||y_r||^2_r + ||e_r||^2_r$$

Finally, upon substituting the above into the functional J_s we obtain

$$J'_s = ||y'_s||^2_s + ||e'_s||^2_s = ||\underline{E}^r y'_s||^2_s + E^r e'_s||^2_s + ||\underline{E}_r y'_s||^2_s + ||E_r e'_s||^2_s$$

$$= ||\underline{E}^r y_s||^2_s + ||E^r e_s||^2_s + ||\underline{E}_r y'_r||^2_s + ||E_r e'_r||^2_s$$

$$= ||\underline{E}^r y_s||^2_s + ||E^r e_s||^2_s + ||y'_r||^2_r + ||e'_r||^2_r$$

$$< ||\underline{E}^r y_s||^2_s + ||E^r e_s||^2_s + ||\underline{E}_r y_r||^2_s + ||e_r||^2_r$$

$$=||\underline{E}^r y_s||_s^2+||E^r e_s||_s^2+||\underline{E}_r y_r||_s^2+||E_r e_r||_s^2$$

$$=||\underline{E}^r y_s||_s^2+||E^r e_s||_s^2+||\underline{E}_r y_s||_s^2+||E_r e_s||_s^2$$

$$=||y_s||_s^2+||e_s||_s^2 = J_s$$

which contradicts our assumption that e_s minimizes the functional under the given constraints thereby verifying the contention.

3. The Memoryless Controller Theorem

Theorem: For a given state feedback configuration and initial time s let the regulator problem have a solution F_r for each $r \geq s$. Then it also admits a memoryless solution M_r for each $r \geq s$.

Proof: Clearly, it suffices to show that M_s exists since for any other $r > s$ we may take $s' = r$. Since F_s solves the regulator problem

$$e_s = F_s y_s = F_s[P_s e_s + d_s(x_s)]$$

minimizes

$$J_s = ||y_s||_s^2 + ||e_s||_s^2$$

under the constraint that

$$y_s = P_s e_s + d_s(x_s)$$

Now, for any partition

$$s = t_o < t_1 < t_2 < \ldots < t_n = \infty$$

of (s,∞) we may write

$$e_s = \sum_{i=1}^{n} [E^{t_i}-E^{t_{i-1}}]e_s = \sum_{i=1}^{n} [E^{t_i}-E^{t_{i-1}}]E_{t_{i-1}}e_s$$

and by the lemma $E_{t_{i-1}} e_s$ is a solution to the regulator problem with initial time t_{i-1} and initial condition

$$x_{t_{i-1}} = SRC\int_{(s,t_{i-1})} \phi(t_{i-1},q)G(q)dE(q)e_s + \phi(t_{i-1},s)x_s$$

Thus since $F_{t_{i-1}}$ is the optimal controller with initial time t_{i-1} we have

$$E_{t_{i-1}} e_s = F_{t_{i-1}}[P_{t_{i-1}} E_{t_{i-1}} e_s + d_{t_{i-1}}(x_{t_{i-1}})]$$

$$= F_{t_{i-1}}[E_{t_{i-1}} y_s]$$

where the last equality results from an argument similar to that used in the proof of the lemma. We may now write e_s as

$$e_s = \sum_{i=1}^{n}[E^{t_i} - E^{t_{i-1}}]E_{t_{i-1}} e_s + \sum_{i=1}^{n}[E^{t_i} - E^{t_{i-1}}]F_{t_{i-1}} E_{t_{i-1}} y_s$$

$$= \sum_{i=1}^{n}[E^{t_i} - E^{t_{i-1}}]E^{t_i} F_{t_{i-1}} E_{t_{i-1}} y_s$$

$$= \sum_{i=1}^{n}[E^{t_i} - E^{t_{i-1}}]E^{t_i} F_{t_{i-1}}[E^{t_i} E_{t_{i-1}}]y_s$$

$$= \sum_{i=1}^{n}[E^{t_i} - E^{t_{i-1}}]F_{t_{i-1}}[E^{t_i} - E^{t_{i-1}}]y_s$$

where the second from the last equality results from the causality of $F_{t_{i-1}}$. Now this equality is independent of the partition and hence upon taking limits we obtain

$$e_s = SLC\int_{(s,\infty)} dE(t)F_t dE(t)y_s = M_s y_s$$

which yields a memoryless regulator which is well defined on the range of the plant (plus disturbance) and the proof of the theorem is thus complete.

4. Controllability and Observability

Although we have used the concepts of controllability and observability primarily as a mathematical description of the coupling between the input or output of a system and its state space historically these concepts evolved from the regulator problem (KM-4). Here controllability is the condition required to solve the regulator problem over a finite interval (t,s) under the additional constraint that $x(t) = 0$ and observability is the condition required to compute the state from the system output so as to be able to implement a solution to the regulator problem in those cases where the state is not physically measurable. Depending on the manner in which one desires to solve these problems, various alternative modes of controllability and observability arise, the details of which will not be included here.

D. Problems and Discussion

1. Co-State Decomposition

The dual of the state decomposition is the co-state decomposition (DE-1) wherein the role of past and future are reversed from that of the state decomposition. We say that $(S,G(t),H(t))$ is a co-state decomposition of T if

i) $G(t)$ maps H into S and satisfies the equality $G(t)=G(t)E_t$.

ii) $H(t)$ maps S into H and satisfies the equality $H(t)=E^tH(t)$.

iii) $G(t)H(t)=E^tTE_t$.

In essence the co-state variable $x(t)=G(t)u$ describes the effect of future inputs on past outputs. As in the case of the state decomposition we say that $(S,G(t),H(t))$ is minimal if $G(t)$ is onto and $H(t)$ is one-to-one for all t.

Problem: Show that every linear operator on (H,E) has a co-state decomposition.

Problem: Give an example of an operator that does not have a minimal co-state decomposition.

Problem: Show that the minimal co-state decomposition of an operator is unique (in a sense similar to that used for the uniqueness of the state decomposition) if it exists.

Problem: Show that an operator has a zero minimal co-state decomposition if and only if it is causal.

Following our pattern of duality we may define co-transition operators via the commutativity of the diagram

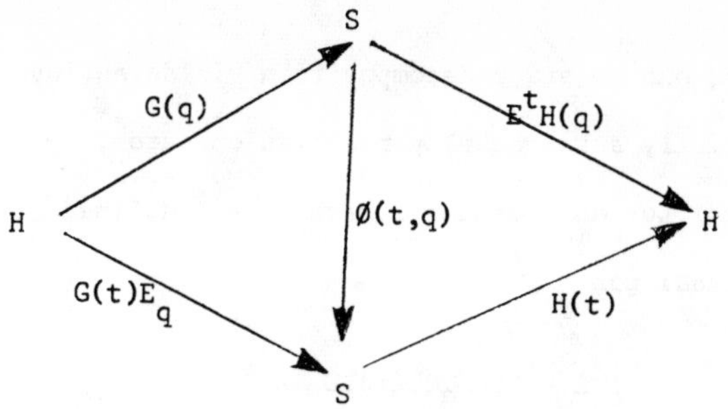

for $t \leq q$.

Problem: Show that every minimal co-state decomposition admits a unique family of co-transition operators for all $t \leq q$.

In general any family of operators rendering the above diagram commutative for all $t \leq q$ are termed co-transition operators.

Problem: Show that a family of co-transition operators satisfies

$$\emptyset(t,t) = I$$

and

$$\emptyset(t,q)\emptyset(q,r) = \emptyset(t,r)$$

whenever $t \leq q \leq r$.

Problem: Let T be a linear operator on (H,E) with co-state decomposition (S,G(t),H(t)) and co-transition operators $\emptyset(t,q)$. Then for any input u and $t \leq s$ in G show that

$$x(t) = SLC\int_{(t,s)} \emptyset(t,q)G(q)dE(q)u + \emptyset(t,s)x(s)$$

Problem: Let T be a linear operator on (H,E) with co-state decomposition (S,G(t),H(t)) and co-transition operators $\emptyset(t,q)$. Then for any input u in H and t in G show that

$$x(t) = SLC\int_{(t,\infty)} \emptyset(t,q)G(q)dE(q)u$$

Finally, our co-state decomposition yields an invariant for the strongly strictly anti-causal part of an operator.

Problem: Let an operator T have a well defined strongly strictly anti-causal part

$$A = SRC\int_G dE(t)TE_t$$

and show that any co-minimal co-state decomposition $(S,G(t),H(t))$ for T is a complete set of invariants for A.

Since the co-state decomposition characterizes the strongly strictly anti-causal part of an operator and the state decomposition characterizes the strongly strictly causal part it is natural to conjecture:

Conjecture: There exists a two-sided state decomposition theory which yields a complete set of invariants for an entire operator (less its memoryless part ?).

2. Nonlinear State Decomposition

Our state decomposition concept carries over naturally to the nonlinear case (DE-1, DE-8, DE-10) though in this instance the state space has two components, one describing the effect of past inputs on future outputs and the other describing the effect of past inputs on the future of the system. Given a nonlinear operator T a state decomposition is a triple $(S,G(t),H(t))$ where

i) S is the product of two Hilbert spaces $S = S_1 \oplus S_2$.

ii) $G(t) = (G_1(t),G_2(t))$ maps H into $S = S_1 \oplus S_2$ and satisfies $G(t) = G(t)E^t$.

iii) $H(t) = (H_1(t),H_2(t))$ maps $S = S_1 \oplus S_2$ into $H \oplus L_t$ where L_t is the space of operators on H satisfying $KE_t=E_tK$ and $H_1(t)=E_tH_1(t)$.

iv) $H_1(t)G_1(t) = E_t T E^t$

v) For any two elements u and v in H

$$[H_2(t)(G_2(t)x)]y = T(E^t x + E_t y) - T(E^t x) - T(E_t y)$$

We note that the first component of our nonlinear state decomposition satisfies the same axioms as the linear state decomposition while the second component, which drops out in the linear case, characterizes the effects of past inputs on the system itself.

Although apparently more complicated the nonlinear state decomposition has essentially the same properties as the linear state decomposition. As in that case we say that the state decomposition is <u>minimal</u> if both components of G(t) are onto and both components of H(t) are one-to-one.

<u>Problem</u>: Show that if a minimal state decomposition exists for a nonlinear operator T then it is unique (in the same sense as used in the linear case).

<u>Problem</u>: Let (S,G(t),H(t)) be a minimal state decomposition of a nonlinear operator T and show that there exists a family of transition operators $\emptyset(t,q)$ for all $t \geq q$ which render the diagram

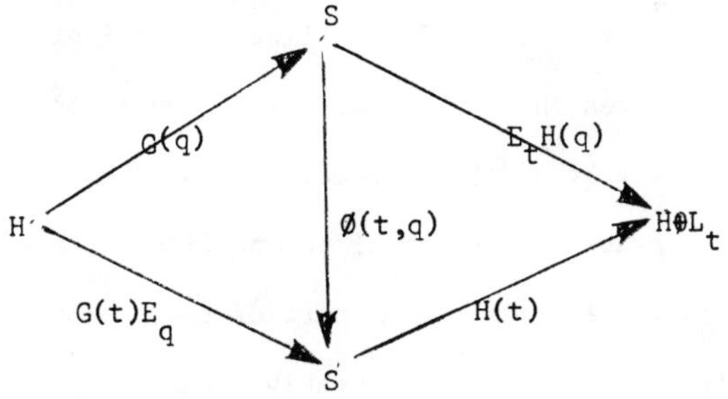

commutative.

As in the linear case we term any family of operators which renders the above diagram commutative <u>transition operators</u> for (S,G(t),H(t)) even if it is not minimal.

Problem: Let (S,G(t),H(t)) be a state decomposition for T with transition operators $\emptyset(t,q)$. Then

$$\emptyset(t,t) = I$$

$$\emptyset(t,q)\emptyset(q,r)=\emptyset(t,r) \text{ for } r \leq q \leq t$$

As in the linear case the concept of state is closely related to causality.

Problem: A nonlinear operator T has a zero state decomposition if and only if it is anti-causal.

Problem: Show that if T is cross-causal (1.G.2) it has a state decomposition with a zero first component.

Problem: Show that if T has a state decomposition with a zero first component that it is the sum of an anti-causal and cross-causal operator.

Consistent with the above we see that the second component of the nonlinear state decomposition corresponds to the cross-causal part of T (1.G.2) whereas results analagous to those in the linear case imply that the first component characterizes the causal part of the operator.

Since a weakly additive operator (1.G.3) contains no cross-causal part and the second component of the nonlinear state decomposition characterizes only the cross-causal part of the operator, we may show:

Problem: Show that every weakly additive operator has a state decomposition with a zero second component.

Since the first component in the nonlinear state decomposition is the same as that in the linear case and the second part may be taken to be zero in the weakly additive case we may study the state concept for weakly additive operators via essentially linear techniques.

Problem: Show that if a weakly additive operator T has state decomposition (S,G(t),H(t)) and transition operators $\emptyset(t,q)$ then for any $t \geq s$ and input u

$$x(t) = SRC\int_{(s,t)}\emptyset(t,q)G(q)dE(q)u+\emptyset(t,q)x(s)$$

Problem: Show that if a weakly additive operator T has a state decomposition (S,G(t),H(t)) and transition operators $\emptyset(t,q)$ then for any t in G and input u in H

$$x(t) = SRC\int_{(-\infty,t)}\emptyset(t,q)G(q)dE(q)u$$

3. The Principle of Optimality

The principle of optimality which we proved as a lemma in our special linear case is fundamental to many optimization problems and is often mistakenly taken as an axiom. In fact, there are a number of practical optimization problems where it is false; but it is true in considerably greater generality than the linear case which we have thus far studied. For this purpose we let (H_s,E), $(\underline{H}_s,\underline{E})$, etc. be as in the regulator problem, P_s be the operator mapping (H_s,E) into $(\underline{H}_s,\underline{E})$ defined by

$$x(t) = SRC\int_{(-\infty,t)}\emptyset(t,q)G(q)dE(q)u$$

where (S,G(t),H(t)) is the state decomposition of a weakly additive operator T with transition operators $\emptyset(t,q)$ and we let

$$d_s(x_s)(t) = \emptyset(t,s)x_s$$

Problem: Let $r \geq s$ and e_s in H_s minimize a weakly additive functional on H_s

$$J_s(y_s,e_s)$$

under the constraint

$$y_s = P_s e_s + d_s(x_s)$$

for some x_s in S. Then show that $e_r = E_r e_s$ minimizes the functional

$$J_r(y_r,e_r) = J_s(E_r y_r,E_r e_r)$$

under the constraint

$$y_r = P_r e_r + d_r(x_r)$$

where x_r in S is defined by

$$x_r = SRC\int_{(s,r)} \emptyset(r,q)G(q)dE(q)e_s + \emptyset(r,s)x_s$$

Problem: Show that the above contention may be false if J_s is not weakly additive.

We note that weakly additive functionals are sometimes called separable functionals (SA-5).

4. TIME INVARIANCE

Unlike the concepts of causality, stability, state, etc. which are well defined in an arbitrary resolution space, time-invariance and the various representations for such operators are well defined only in a restricted class of resolution spaces termed uniform resolution spaces. The purpose of this chapter is the formulation of the uniform resolution space concept and the study of time-invariant operators defined on such spaces. For the most part, an assumption of time-invariance yields little in strengthening the system theoretic results we have already obtained, but it allows a number of powerful operator representation theorems to be invoked for the simplified description of the various properties we have already studied. As such, the chapter is devoted primarily to the formulation of appropriate Fourier and Laplace transformations for time-invariant operators with their explicit interpretation in a systems context left to the problems of the last section.

Our approach to the problem of operator transformation draws on the work of Falb and Freedman (FA-1, FA-2, FE-2), Masani and Rosenberg (MS-1, MS-3, MS-4, RO-1) as unified by the author (SA-2, SA-6, SA-7, SA-8). We begin with the definition and elementary properties of the uniform resolution space and its character space. We then define the basic classes of time-invariant operators encountered in uniform resolution space. This being followed by the development of the Fourier transform both from an abstract spectral theoretic point of view and as a Fourier-Stieljes transform of an appropriate measure associated with an operator. Finally, the Fourier transformation developed for time-invariant operators is combined with Mackey's theory of Laplace transforms on groups (MA-2) to obtain a Laplace transformation for operators.

A. Uniform Resolution Space

1. Definition and Examples

We term a 4-tuple (H,G,E,U) a uniform resolution space if (H,G,E) is a resolution space and U is a group of unitary operators in H, defined for all t in G, satisfying the imprimativity equality (MA-1, MS-3)

$$U^tE(A) = E(A+t)U^t$$

for all Borel sets A of G and t in G. As with a regular resolution space we normally drop the G in our notation and work with a fixed, but arbitrary, G throughout the remainder of the work. Unless specified to the contrary, all operators in the present chapter map a uniform resolution space (H,E,U) into itself. Of course, all of our causality theory applies to such operators when interpreted as operating on the resolution space (H,E).

The classical example of a uniform resolution space which serves as a model for much of our theory is $L_2(G,K,m)$, K a Hilbert space and m the Haar measure on G, together with its usual resolution structure, and the group of shift operators, U, on $L_2(G,K,m)$ defined by

$$(U^tf)(s) = f(s-t)$$

Here

$$(U^tE(a)f)(s)=(U^t\chi_Af)(s)=(\chi_Af)(s-t)=\chi_A(s-t)f(s-t)$$

$$=\chi_{A+t}(s)f(s-t)=\chi_{A+t}(s)(U^tf)(s)=(E(A+t)U^tf)(s)$$

hence the required equality is verified and we do, indeed, have a uniform resolution space. Note that if we replace m by any measure other than the Haar measure, the shift operators are no longer unitary, in fact they might not even be well defined (SA-2), and the above space is no longer uniform. Of course, it may be possible to use a weighted shift to define a uniform structure for $L_2(G,K,\mu)$.

In the sequel we say that the space $L_2(G,K,m)$ with its usual resolution structure and the group of shift operators taken for U has the usual uniform resolution structure.

2. The Character Space

Given any uniform resolution space, (H,G,E,U) Stone's theorem (C.A.2) allows one to uniquely construct a second spectral measure and unitary group defined on G_T, the group of ordinary characters, via

$$\hat{U}^{\gamma} = L\int_G (\gamma,-t)_T dE(t)$$

and

$$U^t = L\int_{G_T} (\gamma,-t)_T d\hat{E}(\gamma)$$

where the Lebesgue integrals are well defined since spectral measures have finite semi-variation relative to scaler valued functions (B.A.2). We term the 4-tuple $(H,G_T,\hat{E},\hat{U})$ the character space for (H,G,E,U) and denote it by $(H,\hat{E},\hat{U})$ when G, and hence G_T, is fixed.

Note that the character space fails to be a resolution space since G_T is not, in general, ordered and even when it is orderable it has no ordering naturally induced by that of G (A.C.3). As such, one cannot define causality like concepts in $(H,G_T,\hat{E},\hat{U})$ though symmetric concepts such as memorylessness have "duals" in the character space.

Although $(H,G_T,\hat{E},\hat{U})$ is not a resolution space it is characterized by many of the same phenomena. In particular, an equality similar to that constraining E and U holds for $\hat{E}$ and $\hat{U}$. Moreover, it is possible to construct a relationship relating a uniform resolution space and its character space independently of Stone's theorem (MA-1).

Theorem: Let (H,E,U) be a uniform resolution space with character space $(H,\hat{E},\hat{U})$. Then for any Borel set, A, of G_T, γ in G_T and t in G the following equalities hold.

i) $$U^t\hat{U}^\gamma = (\gamma, t)_T\hat{U}^\gamma U^t$$

ii) $$\hat{U}^\gamma E(A) = E(A-\gamma)\hat{U}^\gamma$$

Proof: i). By Stone's theorem

$$U^t\hat{U}^\gamma = U^t[L\int_G(\gamma,-q)_T dE(q)]=[L\int_G(\gamma,-q)_T dE(q+t)]U^t$$

$$=[L\int_G(\gamma, t)_T(\gamma,-q-t)_T dE(q+t)]U^t = (\gamma, t)_T\hat{U}^\gamma U^t$$

Here the second equality is due to the defining equality for U^t and all of the integrals involved are assured to exist since ordinary characters are totally measurable.

ii). Denote by V the unitary on G defined by $V^t = (-\gamma,t)_T U^t$ for a fixed γ, and denote by E_1 and E_2 the spectral measures defined by

$$\hat{E}_1(A) = E(A+\gamma)$$

and

$$\hat{E}_2(A) = \hat{U}^{-\gamma}\hat{E}(A)\hat{U}^\gamma$$

Now

$$V^t=(-\gamma,-t)_T U^t=(-\gamma,-t)_T[L\int_{G_T}(\theta,-t)_T d\hat{E}(\theta)]$$

$$=L\int_{G_T}(\theta-\gamma,-t)_T d\hat{E}(\theta)= =L\int_{G_T}(\emptyset,-t)_T d\hat{E}(\emptyset+\gamma)$$

$$= L\int_{G_T} (\emptyset,-t)_T d\hat{E}_1(\emptyset)$$

while

$$V^t = (-\gamma,-t)_T U^t = (-\gamma,-t)_T \hat{U}^{-\gamma}\hat{U}^{\gamma}U^t = \hat{U}^{-\gamma}(-\gamma,-t)_T \hat{U}^{\gamma}U^t$$

$$= \hat{U}^{-\gamma}U^t\hat{U}^{\gamma} = \hat{U}^{-\gamma}[L\hat{\ }_{G_T}(\emptyset,-t)_T dE(\emptyset)]\hat{U}^{\gamma}$$

$$= L\int_{G_T}(\emptyset,-t)_T \hat{U}^{-\gamma} dE(\emptyset)\hat{U}^{\gamma} = L\int_{G_T}(\emptyset-t)_T d\hat{E}_2(\emptyset)$$

where, once again, all of the integrals are assured to exist since characters are totally measurable. Now we have two representations of V^t hence by the uniqueness of Stone's theorem $E_1 = E_2$ or equivalently

$$\hat{E}(A+\gamma) = \hat{U}^{-\gamma}E(A)\hat{U}^{\gamma}$$

for each γ in G_T and Borel set A which is the desired equality after multiplying through by $\hat{U}^{\gamma}$. The proof of the theorem is therefore complete.

For future reference the various equalities relating E,E,U and $\hat{U}$ are tabulated below.

i) $U^tE(A) = E(A+t)U^t$ for any t in G and Borel set A of G.

ii) $U^tE^s = E^{s+t}U^t$ for any t and s in G.

iii) $U^tE_s = E_{s+t}U^t$ for any t and s in G.

iv) $\hat{U}^{\gamma}\hat{E}(A) = \hat{E}(A-\gamma)\hat{U}^{\gamma}$ for any γ in G and Borel set A of G_T.

v) $U^t\hat{U}^{\gamma} = (\gamma, t)_T\hat{U}^{\gamma}U^t$ for any t in G and γ in G_T.

B. Spaces of Time-Invariant Operators

1. Time-Invariant Operators

Since G_T is not ordered there is no analog of the causal operators on the character space. We do, however, have an analog of the memoryless operators defined on the character space, the time-invariant operators, since memorylessness is a symmetric concept defined independently of the ordering on G. We say that an operator T on H is time-invariant if

$$\hat{E}(A)T = T\hat{E}(A)$$

for all Borel sets A of G_T. Standard spectral theoretic results yield:

Theorem: For a linear bounded operator, T, the following are equivalent.

i) T is time-invariant.

ii) $U^t T = TU^t$ for all t in G.

iii) $H(A) = Im(E(A))$ is an invariant subspace of T for all Borel sets A of G_T.

Consider the operator, T, on $L_2(Z,R,m)$ with its usual uniform resolution structure defined by

$$y(k) = \sum_{j=-\infty}^{\infty} \underline{T}(k,j)u(j)$$

Now for any i in Z and u in $L_2(Z,R,m)$

$$(U^i Tu)(k) = y(k-i) = \sum_{j=-\infty}^{\infty} \underline{T}(k-i,j)u(j)$$

and

$$(TU^i u)(k) = \sum_{j=-\infty}^{\infty} \underline{T}(k,j)u(j-i) = \sum_{n=-\infty}^{\infty} \underline{T}(k,n+i)u(n)$$

thus if equality is to hold for all u we require that

$$\underline{T}(k-i,j) = \underline{T}(k,j+i)$$

for all k, j and i; or equivalently that

$$\underline{T}(k,j) = \overline{T}(k-j)$$

for some sequence $\overline{T}$ if T is to be time-invariant. Of course, upon identifying $\underline{T}(k,j)$ with the entries of an infinite matrix we see that the matrix represents a time-invariant operator if and only if it is Toeplitz.

Similarly, on $L_2(R,K,m)$ an operator represented by

$$y(t) = \int_G \underline{T}(t,q)u(q)dm(q)$$

is time-invariant if and only if $\underline{T}(t,q) = \overline{T}(t-q)$ for some function, $\overline{T}$, defined on R.

Like their memoryless analogs the time-invariant operators are mathematically well behaved. Standard spectral theoretic arguments yielding:

Theorem: The time-invariant operators on a uniform resolution space, (H,E,U), form a Banach algebra closed in the strong (hence also uniform) operator topology of the algebra of all bounded linear operators on H. Moreover, the adjoint of a time-invariant operator is time-invariant and the inverse of a time-invariant operator, if it exists, is time-invariant.

2. Static Operators

On a uniform resolution space, not only is it possible to define operator classes on the character space which are "dual" to those defined on the resolution space itself, but we may also define operators whose properties are determined simultaneously by the given uniform resolution space and its character space. Our first such class of operators are the static operators which are required to be both memoryless and time-invariant. These, in some sense, serving as the "scalars" of our theory. Clearly, since the static operators are the intersection of two Banach algebras we have:

Theorem: The bounded static operators on a uniform resolution space (H,E,U) form a Banach algebra closed in the strong (hence also uniform) operator topology of the algebra of all bounded linear operators on H. Moreover, the adjoint of a static operator is static and the inverse of a static operator, if it exists, is static.

The operator represented by

$$y(k) = \sum_{j=-\infty}^{\infty} \underline{T}(k,j)u(j)$$

on $L_2(Z,R,m)$ with its usual uniform resolution structure is memoryless if and only if $\underline{T}(k,j) = 0$ for $k \neq j$ and time invariant if and only if $\underline{T}(k,j) = \overline{T}(k-j)$ for some sequence $\overline{T}$. Upon combining these two constraints we thus find that the operator is static if and only if

$$\underline{T}(k,j) = \begin{cases} \overline{T}(0) & k = j \\ 0 & k \neq j \end{cases}$$

hence the operator is actually in the form

$$y(k) = \overline{T}(0)u(k)$$

that is, a constant multiplier.

Similarly, on $L_2(G,K,m)$, K a Hilbert space, the operator defined by

$$y(t) = \overline{T}(0)u(t)$$

where $\overline{T}(0)$ is an operator on K, is static and all static convolution operators are of this form.

3. Subordinative Operators

Another class of time-invariant operators which is defined on a uniform resolution space by making recourse to both the primary and character structure are the trigonometric polynomials. That is, operators of the form

$$\sum_{i=1}^{n} R_i U^{t_i}$$

where R_i is static, the t_i are arbitrarily chosen elements of G and n is arbitrary. The use of the term trigonometric polynomial for this class of operators will become more apparent when their Fourier Transform is formulated in the following section. Clearly, the triganometric polynomials are closed under addition, multiplication, and adjoints while since each such polynomial is a linear combination of products of time-invariant operators, they are time-invariant.

Theorem: The bounded trigonometric polynomials form a sub-algebra of the time-invariant operators. Moreover, the adjoint of a triganometric polynomial is also a trigonometric polynomial.

Since the trigonometric polynomials are not closed it is natural to define the subordinative operators as the closure in the uniform operator topology of the triganometric polynomials (MS-1, KG-1). Clearly, it follows immediately from their definition that:

Theorem: The subordinative operators on a uniform resolution space, (H,E,U), form a Banach algebra closed in the uniform operator topology of the Banach algebra of all time-invariant operators on H.

Moreover, the adjoint of a subordinative operator is also subordinative.

One final class of operators which proves to be important in our theory of time-invariant operators are the simple operators

$$\sum_{i=1}^{n} R_i \hat{E}(A_i)$$

where R_i is static, A_i is an open set in G_T for all i, and n is arbitrary. As with the trigonometric polynomials the source of the term "simple operators" will become apparent when the Fourier transformation of such an operator is formulated. As with the trigonometric polynomials we have:

Theorem: The bounded simple operators form a sub-algebra of the time-invariant operators. Moreover, the adjoint of a simple operator is simple.

Since the simple operators do not form a closed subspace in the time-invariant operators it is natural, as in the case of the trigonometric polynomials, to define a new class of operators as their uniform closure. In fact, this results in precisely the subordinative operators (MS-1, SA-7).

Theorem: The simple operators form a dense subset of the subordinative operators in the uniform operator topology.

Proof: Since for each open B in G_T, $\hat{E}(B)$ is the limit of a sequence of triganometric polynomials each such $\hat{E}(B)$ is indeed subordinative (C.A.2). On the other hand

$$U^t = L\int_{G_T} (\gamma,-t)_T d\hat{E}(\gamma)$$

where $(\gamma,-t)_T$ is in $R_\infty(G_T,R,\hat{E})$ since it is continuous and totally integrable. As such, it is approximatable in the $|| \ ||_\infty$ norm by a sequence of step functions

$$S^k = \sum_{i=1}^{n(k)} c_i^k \chi_{A_i}$$

where the c_i^k are scalars and the A_i are open sets in G_T. The definition of the Lebesgue integral thus implies that

$$U^t = \underset{k \to \infty}{\text{limit}} \sum_{i=1}^{n(k)} c_i^k \hat{E}(A_i)$$

showing that each U^t is in the closure of the simple operators in the uniform operator topology. Similarly, for any static operator R

$$RU^t = \underset{k \quad \infty}{\text{limit}} \sum_{i=1}^{n(k)} c_k^i RE(A_i)$$

is in the closure of the simple operators. Finally, the linear combinations of such operators, the trigonometric polynomials, are in this space (since it is linear) and the limits of such linear combinations, and subordinative operators, are in this space (since it is closed). As such, the subordinative operators are contained in the (uniform) closure of the simple operators which combined with the fact that the simple operators are subordinative (since $\hat{E}(B)$ is subordinative for any open set B of G_T) yields the desired result.

4. Operator Space Relationships

In general there is no immediate relationship between time-invariance and causality. For instance the shift, U^s, satisfies

$$E^tU^s = U^sE^{t-s} = U^sE^{t-s}E^t = E^tU^sE^t$$

if and only if $s \geq 0$. Thus the shift is causal if and only if $s \geq 0$ and by a similar argument anti-causal if and only if $s \leq 0$. Although time-invariance does not imply causality or anticausality an assumption of time-invariance often simplifies the determination of causality (SA-7).

Theorem: Let T be bounded time-invariant operator. Then T is causal if and only if $E^tT = E^tTE^t$ for some t in G.

Proof: Clearly, causality implies that $E^tT = E^tTE^t$ for all t in G; hence the proof is trivial in one direction. On the other hand, if

$$E^{t_o}T = E^{t_o}TE^{t_o}$$

for some t_o then

$$E^tT = E^tU^{(t-t_o)}U^{-(t-t_o)}T = U^{(t-t_o)}E^{t_o}TU^{0(t-t_o)}$$

$$= U^{(t-t_o)}E^{t_o}TE^{t_o}U^{-(t-t_o)} = E^tU^{(t-t_o)}U^{-(t-t_o)}TE^t$$

$$= E^tTE^t$$

for all t; hence T is causal, and the proof is complete. Of course, a similar condition applies to all of our other equivalent conditions for causality (SA-4, PR-4) where the "for all t in G" can be replaced with "for some t in G" and similarly for the other causality like conditions.

Similarly, a trigonometric polynomial

$$\sum_{i=1}^{n} R_i U^{t_i}$$

is causal if $t_i \geq 0$ for all i and the uniform limit of such triganometric polynomials is also causal (since the causal operators are closed in the uniform operator topology). Of course, all static operators are memoryless (hence both causal and anti-causal). Finally, we note that the simple operators intersect with the causals only in the zero operator though, of course, their closure, the subordinatives, contains many causal operators.

Consistent with the rather large variety of time-invariant operator classes with which we deal it is convenient to adopt special notation to denote them as follows.

i) $\mathcal{B}$ = bounded operators on H.

ii) $\mathcal{T}$ = time-invariant operators.

iii) $\mathcal{R}$ = static operators.

iv) $\mathcal{S}$ = subordinative operators

v) $\mathcal{P}$ = triganometric polynomials.

vi) $\mathcal{Q}$ = simple operators.

The relationship between the above operator classes is described by the following diagram.

$$\begin{array}{ccccccccc} & \subset & \mathcal{P} & \subset & & & & & \\ \mathcal{R} & & & & \mathcal{S} & \subset & \mathcal{T} & \subset & \mathcal{B} \\ & \subset & \mathcal{Q} & \subset & & & & & \end{array}$$

C. The Fourier Transform

1. Definition and Examples

For an operator, T, on a uniform resolution space, (H,E,U), we term a static operator valued Riemann totally integrable function, $\hat{T}$ in $R_\infty(G_T,R,\hat{E})$, its Fourier Transform if

$$T = L\int_{G_T} \hat{T}(\gamma)d\hat{E}(\gamma)$$

For a static operator, R, we have

$$R=RI=R[L\int_{G_T} d\hat{E}(\gamma)]=L\int_{G_T} Rd\hat{E}(\gamma)=L\int_{G_T} \hat{R}(\gamma)d\hat{E}(\gamma)$$

where $\hat{R}$ is the constant function equal to R for all γ in G_T. Thus the Fourier transform of R is $\hat{R}$. For a shift operator, U^t, Stone's theorem yields

$$U^t = L\int_{G_T} (\gamma,-t)_T d\hat{E}(\gamma)$$

hence the Fourier transform of a shift is the character $(\gamma,-t)_T$ (times the identity operator). Combining these two results we obtain as the Fourier transform of a trigonometric polynomial

$$\sum_{i=1}^{n} R_i U^{t_i}$$

the function

$$\sum_{i=1}^{n} R_i(\gamma,-t_i)_T$$

which is a trigonometric polynomial in the classical sense, though with static operator coefficients instead of scalers, and the source of our terminology for this class of operators. Similarly, the Fourier transform of a simple operator

$$\sum_{i=1}^{n} R_i\hat{E}(A_i)$$

is the simple function with static operator coefficients

$$\sum_{i=1}^{n} R_i\chi_{A_i}(\gamma)$$

2. An Isomorphism

The following lemma which illustrates the "scalar like" characteristics of the static operators proves to be of fundamental importance in our characterization theorem for the Fourier transform (SA-7).

Lemma: Let R be an arbitrary static operator and A be a non-void open set in G_T. Then

$$||R\hat{E}(A)|| = ||R||$$

Proof: Since $\hat{U}^{-\gamma}$ is unitary, $\hat{U}^{\gamma}\hat{E}(A) = \hat{E}(A-\gamma)\hat{U}^{\gamma}$ and R is memoryless

$$||R\hat{E}(A)||=||\hat{U}^{-\gamma}R\hat{E}(A)||=||R\hat{U}^{-\gamma}\hat{E}(A)||=||R\hat{E}(A+\gamma)\hat{U}^{-\gamma}||=||R\hat{E}(A+\gamma)||$$

for all γ in G_T. Now since A is open, the sets $A+\gamma$ for all γ in G_T form an open covering of G_T and since G_T is Lindelof (i.e., since G is ordered G_T is either a connected compact set or the product of R and a connected compact set (A.C.2) and since both of these are second countable G_T is second countable and hence Lindelof (KE-1)) there exists a countable family of characters; γ_i, i=1,2,...; such that the sets $A+\gamma_i$ cover G_T. Now defining a family of sets A_n via

$$A_n = A+\gamma_n \setminus \bigcup_{i=1}^{n-1} A_i$$

we obtain a disjoint family of Borel sets which cover G_T and satisfy

$$||R\hat{E}(A_n)||=||R\hat{E}(A_n)\hat{E}(A+\gamma_n)||\leq||R\hat{E}(A+\gamma_n)||=||R\hat{E}(A)||$$

with equality for n = 1. Finally since $\hat{E}$ is countably additive

$$||R\hat{E}(A)||\leq||R||\ ||\hat{E}(A)||$$

$$\leq||R||=||R[\sum_{i=1}^{\infty}\hat{E}(A_i)]||=||\underset{n\to\infty}{\text{s-limit}}\ R\ [\sum_{i=1}^{n}\hat{E}(A_i)]||$$

$$\leq\underset{n\to\infty}{\text{limit}}||R[\sum_{i=1}^{n}\hat{E}(A_i)]||=\underset{n\to\infty}{\text{limit}}[\sup_{1\leq i\leq n}||R\hat{E}(A_i)||]$$

$$=||R\hat{E}(A)||$$

Here the limit commutes with R and the norm since both of these are continuous mappings and the sup formula (1.E.3) is valid since R is time-invariant. Finally, the preceeding inequality with equality for n = 1 assures that the sup is equal to $||RE(A)||$ as is to be shown.

With the aid of the lemma we may now prove our main theorem on Fourier transforms.

Theorem: The Fourier transformation is a Banach Algebra isomorphism mapping $R_\infty(G_T, \mathcal{R}, \hat{E})$ onto S.

Proof: Since $R_\infty(G_T, \mathcal{R}, \hat{E})$ is contained in $L_\infty(G_T, \mathcal{R}, \hat{E})$ the integral is a well defined operator on this space with norm less than or equal to one since $\hat{E}$ has semi-variation one with respect to time-invariant operators (since they commute with $\hat{E}$) and hence also static operators (B.A.2). Of course, the integral defines a linear operator and since $\hat{E}$ is idempotent it is multiplicative (B.D.2). Moreover, if

$$T = L\int_{G_T}\hat{T}(\gamma)d\hat{E}(\gamma)$$

where T is in $R_\infty(G_T, \mathcal{R}, \hat{E})$ then T can be approximated in $|| \; ||_\infty$ by a sequence of step functions

$$S^k = \sum_{i=1}^{n(k)} R_i^k \chi_{A_i^k}$$

where A_i^k is open and R_i^k is static. As such,

$$T = \lim_{k \to \infty} R_i^k \hat{E}(A_i^k)$$

and is thus subordinative. The integral thus maps $R_\infty(G_T, \mathcal{R}, \hat{E})$ into $\mathcal{S}$. To complete the proof it thus remains to show that the integral, in fact, maps $R_\infty(G_T, \mathcal{R}, \hat{E})$ onto $\mathcal{S}$ and that it has norm greater than or equal to one. In the latter case it suffices to consider simple operators

$$T = \sum_{i=1}^{n} R_i \hat{E}(A_i) = L\int_{G_T} \hat{T}(\gamma) d\hat{E}(\gamma) = L\int_{G_T} [\sum_{i=1}^{n} R_i \chi_{A_i}(\gamma)] d\hat{E}(\gamma)$$

R_i static and A_i open since such operators are dense in $\mathcal{S}$ and the integral is a continuous linear operator (since its norm is less than or equal to one). Now for $1 \leq j \leq n$ the lemma implies that

$$||T|| \geq ||TE(A_j)|| = ||\sum_{i=1}^{n} R_i E(A_i) E(A_j)|| = ||R_j E(A_j)|| = ||R_j||$$

hence upon taking the sup over all j we have

$$||T|| \geq \sup_j ||R_j|| = ||\hat{T}||_\infty$$

verifying that the norm of the integral operator is greater than or equal to one which together with the opposite inequality which has already been verified assures that the Fourier transform is an isometric operation.

Finally, if T is subordinative it can be expressed as the (uniform) limit of simple operators

$$T = \lim_{k \to \infty} S^k = \lim_{k \to \infty} \left[\sum_{i=1}^{n(k)} R_i^k \hat{E}(A_i^k) \right]$$

where each S^k has the Fourier transformation

$$\hat{S}^k = \sum_{i=1}^{n(k)} R_i^k \chi_{A_i^k}$$

Now, the S^k converge (uniformly) to T and hence must be a Cauchy sequence; hence, the isometric character of the Fourier transformation assures that the $\hat{S}^k$ also form a Cauchy sequence (in $|| \; ||_\infty$) and hence, since $R_\infty(G_T, \mathcal{R}, \hat{E})$ is complete they converge to a function, $\hat{T}$, in $R_\infty(G_T, \mathcal{R}, \hat{E})$ which by the continuity of the integral operator must be the Fourier transform of T. The integral operator thus maps onto S and the proof of the theorem is complete.

The theorem leads immediately to the following list of classical properties dealing with the relationship of an operator and its Fourier transform.

i) $\widehat{T+S} = \hat{T} + \hat{S}$ for any operators T and S in $\mathcal{S}$.

ii) $\widehat{cT} = c\,\hat{T}$ for any T in $\mathcal{S}$ and scaler c.

iii) $\widehat{TS} = \hat{T}\,\hat{S}$ for any operators T and S in $\mathcal{S}$.

iv) $||T|| = ||\hat{T}||_\infty$ for any operator T in $\mathcal{S}$.

v) An operator has a Fourier transform if and only if it is subordinative in which case it is unique a.e.

vi) $\hat{T}^*(\gamma) = \hat{T}(\gamma)^*$ for any operator T in $\mathcal{S}$.

vii) T is a positive operator if and only if $\hat{T}(\gamma)$ is positive for almost all γ in G_T.

viii) $T^{-1}(\gamma) = T(\gamma)^{-1}$ for almost all γ in G_T where the inverses on the right side of the equality exist a.e. if and only if T^{-1} exists and is subordinative for any subordinative operator T in $\mathcal{S}$.

i) through v) are simply statements of the fact that the Fourier transform is a Banach Algebra isomorphism between $R_\infty(G_T,\mathcal{R},\hat{E})$ and $\mathcal{S}$, vi) results from taking adjoints on both sides of the defining equality for the Fourier transform, vii) results from an argument parallel to that used in the proof of the theorem and lemma to prove that the Fourier transformation is an isometry, and viii) follows from iii) in the obvious manner.

3. Weighting Measures

In the preceeding, the Fourier transform was defined as an abstract spectral representation of an operator without recourse to a "Fourier" like integral. In fact, the Fourier transformation of an operator can be interpreted as a Fourier-Stieltjes transform of an appropriate weighting measure representation of an operator (SA-7). The existence of such a representation is predicated on the following lemma.

Lemma: Let $\overline{T}$ be in $M(G,C,\mathcal{R})$, (i.e. $\overline{T}$ is a static operator valued measure defined on the Borel sets of G which has finite semi-variation relative to complex valued functions). Then the integral

$$C\int_G U^t d\overline{T}(t) = T$$

exists and is a subordinative operator. (Here the notation $C\int$ is used to indicate that both $LC\int$ and $RC\int$ exist and are equal).

Proof: For each γ in G_T $(\gamma,-t)_T$ is a bounded continuous complex valued function and hence

$$C\int_G (\gamma,-t)_T d\overline{T}(t) = \underset{p}{\text{limit}} \sum_{i=1}^{n(p)} [(\gamma,-r_i)_T \overline{T}(t_{i-1},t_i)]$$

exists, where the limit is taken over the net of partitions, p, of G into finitely many intervals and $r_i = t_i$ or $r_i = t_{i-1}$. Moreover, since $\overline{T}$ has finite semi-variation, the R valued function of G_T defined by this integral is in $R_\infty(G_T,R,\hat{E})$ and hence upon invoking the isometry property of the Fourier transform as a mapping from $R_\infty(G_T,R,\hat{E})$ onto S we have

$$\begin{aligned} T &= L\int_{G_T}[C\int_G(\gamma-t)_T d\overline{T}(t)]d\hat{E}(\gamma) \\ &= L\int_{G_T}[\lim_p \sum_{i=1}^{n(p)} (\gamma,-r_i)_T\overline{T}(t_{i-1},t_i)]d\hat{E}(\gamma) \\ &= \lim_p \sum_{i=1}^{n(p)} [L\int_{G_T}(\gamma,-r_i)_T d\hat{E}(\gamma)]\overline{T}(t_{i-1},t_i) \\ &= \lim_p \sum_{i=1}^{n(p)} U^{r_i}\overline{T}(t_{i-1},t_i) = C\int_G U^t d\overline{T}(t) \end{aligned}$$

Here the commutation of the limit and the integral is valid since the Fourier transformation is isometric and the existence of the Lebesgue integral of

$$C\int_G(\gamma,-t)_T d\overline{T}(t)$$

is assured since this function is in $R_\infty(G_T, R,\hat{E})$. The required Cauchy integral therefore exists and since its value is the Fourier transform of a function in $R_\infty(G_T,R,\hat{E})$ it is subordinative, thereby completing the proof.

By a <u>weighting measure</u> for an operator, T, we mean a measure, $\overline{T}$, in $M(G,C,R)$ such that

$$T = C\int_G U^t d\overline{T}(t)$$

Clearly, subordinativity is a necessary condition for a weighting measure to exist though it is not sufficient.

For a static operator, R, $\overline{R}$ is a point mass equal to R at zero; whereas for an operator RU^t, R static, $\overline{RU^t}$ is a point mass equal to R at t. Now, by linearity for any trigonometric polynomial

$$P = \sum_{i=1}^{n} R_i U^{t_i}$$

$\overline{P}$ has point masses equal to R_i at t_i. Possibly the most important class of operators which admit weighting measure representations are those characterized by a convolutional integral (FA-1)

$$y(t) = \int_G T(t-q)u(q)dm(q)$$

on the resolution space $L_2(G,K,m)$. Here,

$$\overline{T}(A) = \int_A T(q)dm(q)$$

is the corresponding weighting measure. In essence, the weighting measure representation is the generalization of the convolution to resolution spaces which do not have a natural function space representation, though in fact it is slightly more general than the convolution on $L_2(G,K,m)$.

4. Fourier-Stieljes Transforms

In the process of proving the lemma (4.C.3) we developed the equality

$$L\int_{G_T} [C\int_C (\gamma,-t)_T d\overline{T}(t)]d\hat{E}(\gamma) = T = C\int_C U^t d\overline{T}(t)$$

hence we have formulated our required relationship between the abstract Fourier transform and the classical Fourier Steiljes transform.

Theorem: Let an operator T have weighting measure $\overline{T}$, then the Fourier Transform of T is equal to the Fourier-Steiltjes transform of $\overline{T}$, i.e.

$$\hat{T}(\gamma) = C\int_G (\gamma,-t)_T d\overline{T}(t)$$

In particular, for the case of a $\overline{T}$ associated with a convolution operator we have

$$\hat{T}(\gamma) = C\int_G (\gamma,-t)_T T(t)dm(t)$$

which is the classical Fourier transform of the weighting function $\underline{T}$ (FA-1).

With the aid of our theorem relating the Fourier transform of an operator and its weighting measure, we may, upon invoking the isomorphism theorem for the Fourier transform (4.C.2), obtain the following properties of the weighting measure representation of an operator.

i) $\overline{T+S} = \overline{T} + \overline{S}$ for any two operators T and S with weighting measures $\overline{T}$ and $\overline{S}$.

ii) $\overline{cT} = c\overline{T}$ for any operator T with weighting measure $\overline{T}$ and scalar c.

iii) $||T|| \leq |\overline{T}|_C$ for any operator T with weighting measure $\overline{T}$.

iv) $\overline{TS} = \overline{T}*\overline{S}$ for any operators T and S with weighting measures $\overline{T}$ and $\overline{S}$ where

$$\overline{T}*\overline{S}(A) = C\int_G \overline{T}(A-t)d\overline{S}(t)$$

is the <u>convolution</u> of $\overline{T}$ and $\overline{S}$.

v) $\overline{T^*} = \overline{T}_*$ for any operator T with weighting measure $\overline{T}$ where

$$\overline{T}_*(A) = \overline{T}(-A)^*$$

is the <u>conjugate</u> of $\overline{T}$.

vi) If T has weighting measure $\overline{T}$ then T is causal if and only if $\overline{T}$ has its support in $[0,\infty)$.

i). and ii). result from the linearity of the integral defining the weighting measure, iii) is due to the fact that

$$||T|| = ||\hat{T}||_\infty \leq |\overline{T}|_C$$

since $\hat{T}$ is the Fourier-Stieltjes transform of $\overline{T}$ and $(\gamma,-t)_T$ has uniform norm one. On the other hand iv), v) and vi) follow from the definition of the weighting measure via the "obvious" manipulations.

D. The Laplace Transform

1. Multiplication

To define the Laplace transform of an operator T we define a class of weighted operators, T_σ, σ in G_P, and let the Laplace transform of T at a complex character $z = \sigma + \gamma$, σ in G_P and γ in G_T, be the Fourier transform of T_σ if it is subordinative. To define the appropriate class of weighted operators we require a concept of multiplication of an operator by a function. To this end if f is a continuous complex valued function defined on G and T is an operator then by $f\cdot T$ we mean the operator

$$f\cdot T = C\int_G dE(t)[C\int_G f(t-q)TdE(q)]$$

when it exists. Since the Cauchy integral is in general unbounded, so is the multiplication operation and hence its domain of definition is not readily characterizable. The operator is, however, clearly linear and multiplicative (via the idempotent character of the measures (B.D.2)). i.e.,

$$f\cdot(g\cdot T) = (fg)\cdot T = (gf)\cdot T = g\cdot(f\cdot T)$$

Let R be static and s be in G and consider the operator

$$f\cdot RU^s = C\int_G dE(t)[C\int_G f(t-q)RU^s dE(q)]$$

$$= C\int_G dE(t)[C\int_G dE(q+s)f(t-q)RU^s]$$

$$= C\int_G dE(t)f(s)RU^s = f(s)RU^s$$

Here the first equality is the definition of multiplication by f, the second results from the defining equality for U^s and the third from the idempotent character of the integrals. By linearity for any trigonometric polynomial (4.B.3)

$$P = \sum_{i=1}^{n} R_i U^{t_i}$$

we have

$$f\cdot P = \sum_{i=1}^{n} f(t_i) R_i U^{t_i}$$

hence for any function multiplication of a triganometric polynomial is well defined. Similarly, if T has a weighting measure and multiplication by f is assumed to be well defined then

$$f\cdot T = f\cdot C\int_G U^t d\overline{T}(t) = C\int_G f(t) U^t d\overline{T}(t) = C\int_G U^t d\overline{T}_f(t)$$

where

$$\overline{T}_f(A) = C\int_G f(t) d\overline{T}(t)$$

if all of the integrals involved exist.

Since multiplication by f is always well defined on $\mathcal{P}$ if one can show that it is a bounded operator on $\mathcal{P}$ then it can be extended to $\mathcal{S}$. In general this is not the case, the following lemma, however, (RU-1) characterizes a class of functions for which multiplication by f is a bounded operator on $\mathcal{P}$ and hence well defined on $\mathcal{S}$.

<u>Lemma</u>: Let

$$f(t) = \int_{G_T} (\gamma, -t)\, d\mu_T(\gamma)$$

be the Fourier-Steiljes transform of a finite complex valued measure on G_T. Then multiplication by f is a well defined bounded linear operator mapping $\mathcal{S}$ into $\mathcal{S}$.

Proof: Clearly, it suffices to show that multiplication by f is a bounded linear operator on P since then the usual extension theorems assure that it is well defined on S. For any trigonometric polynomial

$$P = \sum_{i=1}^{n} R_i U^{t_i}$$

we have

$$f \cdot P = \sum_{i=1}^{n} f(t_i) R_i U^{t_i}$$

An application of the Fourier transform thus yields

$$||f \cdot P|| = ||\widehat{f \cdot P}||_\infty = \sup_{\gamma \varepsilon G_T} || \sum_{i=1}^{n} f(t_i)(\gamma,-t_i)_T R_i ||$$

Now, for any γ in G_T we have

$$f(t)(\gamma,-t)_T = C\int_G (\theta,-t)_T d\mu(\theta-\gamma)$$

hence

$$||f \cdot P|| = \sup_{\gamma \varepsilon G_T} || \sum_{i=1}^{n} R_i \; C\int_G (\theta,-t_i) d\mu(\theta-\gamma) ||$$

$$= \sup_{\gamma \varepsilon G_T} || C\int_G \sum_{i=1}^{n} R_i (\theta,-t_i)_T d\mu(\theta-\gamma) ||$$

$$\leq ||\hat{P}||_\infty |\mu|_R = ||P|| \; |\mu|_R$$

where $|\mu|_R$ is the semivariation of the measure μ with respect to static operator valued functions which is finite and independent of f since μ is finite and scalar valued. Consistent with the above inequality multiplication by f is a well defined operation on P with norm less than or equal to $|\mu|_R$ and hence uniquely extends to a well defined operation on S thereby completing the proof.

2. Definition and Examples

Upon invoking the above tools we may now define the Laplace Transform of an operator, T, at a complex character $z = r + \gamma$ to be the Fourier transform of the operator

$$T_r = (r,-t)_P.T$$

if T_r is defined and subordinative. We denote the Laplace transformation at z by $\hat{T}(z) = \hat{T}_r(\gamma)$. The use of the same notation as for the Fourier transform causes no difficulty since if T is subordinative $T_o = T$ and $\hat{T}(o+\gamma) = \hat{T}(\gamma)$. As such the Laplace transform is an extension of the Fourier transform for subordinative operators though, in fact, it may also be well defined for non-subordinative operators (for some z). Clearly, the Laplace transform is not well defined for all complex characters, its domain of definition being of the form $G_T \times D$ where D is the set of real characters for which T_r is subordinative. Clearly, 0 is in D if and only if T is subordinative.

Since for every trigonometric polynomial

$$P = \sum_{i=1}^{n} R_i U^{t_i}$$

$(r,-t)_P.P$ is a well defined trigonometric polynomial given by

$$P_r = (r,-t)_P.P = \sum_{i=1}^{n} (r,-t_i)_P R_i U^{t_i}$$

for all r in G_P the Laplace transform of P is defined for all z in G_{C_*} and given by

$$\hat{T}(z)=T(r+\gamma)= \sum_{i=1}^{n} (r,-t_i)_P(\sigma,-t_i)_T R_i = \sum_{i=1}^{n} (z,-t_i)_{C_*} R_i$$

In general, however, multiplication by $(r,-t)_P$ is an unbounded operation on P and hence the Laplace transform is not everywhere defined for subordinative operators though, of course, it is defined for $r = 0$ where it coincides with the Fourier transform. One class of subordinative operators where the Laplace transform is well behaved are the causal subordinative operators of the form

$$T = \underset{k \to \infty}{\text{limit}} \sum_{i=1}^{n(k)} R_i^k U^{t_i^k}$$

where $t_i^k \geq 0$. Now since all of the t_i^k are greater than or equal to zero for any

$$T_k = \sum_{i=1}^{n(k)} R_i^k U^{t_i^k}$$

we have

$$(r,-t)_P \cdot T_k = \sum_{i=1}^{n(k)} (r,-t_i^k)_P R_i^k U^{t_i^k} = (r,-|t|)_P \cdot T_k$$

Now for $r > 0$ $(r,-|t|)_P$ is positive definite hence by Bochner's theorem (RU-1) it is the Fourier-Stieltjes transform of a finite positive measure and by our lemma defines a bounded linear operator on P hence by the above equality $(r,-t)_P$ defines a bounded linear operator on the causal triganometric polynomials and thus may be extended to the causal subordinative operators. As such, for $r > 0$

$$(r,-t)_P \cdot T = (r,-t)_P \cdot [\underset{k \to \infty}{\text{limit}} T_k] = \underset{k \to \infty}{\text{limit}} [(r,-t)_P \cdot T_k]$$

exists and is subordinative. The Laplace transform of such a causal subordinative operator thus exists for z in $G_T x[0,\infty)$ (it exists at $r = 0$ since the operator is subordinative) and is given by

$$T(z) = \underset{k \to \infty}{\text{limit}} \sum_{i=1}^{n(k)} [(z,-t_i^k)_{C_*} R_i^k]$$

in this region. Finally, if T has a weighting measure representation

$$T = C\int_G U^t d\overline{T}(t)$$

the above argument assures that all of the integrals required exist and the Laplace transform is given by

$$T(z) = C\int_G (z,-t)_{C_*} d\overline{T}(t)$$

when $z = r + \gamma$ with $r \geq 0$, which is the classical Laplace transform integral.

3. Region of Existence

Although we assume that G is ordered this does not induce an ordering on G_T but it does induce an ordering on G_P. As such, we may define intervals in G_P. That is, a set I in G_P is an interval if whenever $r_1 \leq r_2$ are in I then any r such that $r_1 \leq r \leq r_2$ is also in I. This definition includes open, closed, half open, infite, semi-infinite and void intervals. Similarly, upon representing G_{C_*} as a product of G_T and G_P we may define strips in G_{C_*} to be sets of the form $G_T x I$ where I is an interval. In $R_{C_*} = C$ we may identify $R_T = R$ with the imaginary axis and $R_P = R$ with the real axis in which case a strip in R_{C_*} is a strip in the usual sense parallel to the imaginary axis. On the other hand, for $Z_{C_*} = C_*$ we may identify $Z_T = T$ with the unit circle and $Z_P = P$ with a ray; hence the strips in Z_{C_*} are anular regions.

Our main theorem on the existence of the Laplace transform is the following (SA-7, MA-1).

Theorem: For any operator T the region of existence of the Laplace transform is a strip, $G_T x I$, in G_{C_*} where I is a (possibly void) interval in G_P.

Proof: Since we already know that the region of definition for the Laplace transform is of the form $G_T x D$ it suffices to show that D is an interval.

That is, if $r_1 \leq r \leq r_2$ then if T_{r_1} and T_{r_2} are subordinative so is T_r. Clearly, this is the case if $r = r_1$ or $r = r_2$; hence we need only consider r in the open interval (r_1,r_2). Now, since the operation of multiplying an operator by a function is linear and multiplicative (in the function) we have

$$T_r=(r,-t)_p\cdot T=\left[\frac{(r,-t)_p}{(r_1,-t)_p+(r_2,-t)_p}\right]\cdot[(r_1,-t)_p+(r_2,-t)_p]\cdot T$$

$$= f\cdot[T_{r_1} + T_{r_2}]$$

if multiplication of the subordinative operator $T_{r_1}+T_{r_2}$ by

$$f(t) = \frac{(r,-t)_p}{(r_1,-t)_p+(r_2,-t)_p}$$

is well defined. Now for $r_1 < r < r_2$ f is of analytic type (RU-1) and hence the Fourier-Steiljes transform of a finite complex measure. As such, the lemma assures that

$$T_r = (r,-t)_p\cdot T = f\cdot[T_{r_1}+T_{r_2}]$$

exists and is subordinative thereby verifying the contention of the theorem.

If we let T and S be operators with Laplace transforms defined in $G_T \times I_T$ and $G_T \times I_S$ respectively, then it follows immediately from the definition of the transform that:

i) $\widehat{T+S}(z) = \hat{T}(z) + \hat{S}(z)$ for z in $G_T \times (I_T \cap I_S)$.

ii) $\widehat{cT}(z) = c\hat{T}(z)$ for z in $G_T \times I_T$.

iii) $\widehat{TS}(z) = \hat{T}(z)\hat{S}(z)$ for z in $G_T \times (I_T \cap I_S)$.

iv) $\widehat{T^*}(z) = \hat{T}(-r+\gamma)^*$ where $z = r + \gamma$ for z in $G_T \times (-I_T)$.

v) $\widehat{T^{-1}}(z) = \hat{T}(z)^{-1}$ for z in $G_T \times (I_T \cap I_{T^{-1}})$.

vi) If $\hat{T}$ exists in a non-void region $\hat{T}$ is unique and T is time-invariant.

vii) If T is subordinative then $\hat{T}$ exists in a non-void region containing $G_T x[0]$.

i) and ii) are simply statements of the linearity of the defining integrals for the Laplace transformation. iii) and iv) result immediately from the corresponding property of the Fourier transformation upon verifying the equalities

$$T_r S_r = (TS)_r$$

and

$$(T^*)_r = (T_{-r})^*$$

v) follows from iii) in the obvious manner, vi) follows from the uniqueness of the Fourier transformation and multiplication by (r,t), and vii) from the definition of the Laplace transformation.

4. Analytic Transforms

If one requires that the operator T_r have a weighting measure representation rather than simply being subordinative than the resultant analytic Laplace transform which is obtained has considerably more structure than the Laplace transform (SA-7). Clearly, the domain of definition of the analytic Laplace transform is contained in the domain of definition of the Laplace transform and the two transforms coincide in this domain.

Although the analytic Laplace transform is defined on a smaller domain than the Laplace transform it has considerably more structure when it exists.

Theorem: For any operator, T, the region of existence for its analytic Laplace transform is a strip $G_T xJ$ in G_{C_*} which is contained in the region of existence of the Laplace transform. Moreover, if $r_1 < r_2$ are any two points in J then for any z in $G_T x(r_1,r_2)$ the analytic Laplace transform of T is given by

$$\hat{T}(z) = C\int_G \left[\frac{(z,-t)}{(r_1-t)_P+(r_2,-t)_P}\right] d(\overline{T}_{r_1}+\overline{T}_{r_2})(t)$$

and is analytic (A.E.3) in this region.

Proof: Via the same argument as for the Laplace transform, if $r_1 < r < r_2$ then T_r exists and is given by

$$T_r = f\cdot(T_{r_1}+T_{r_2})$$

where

$$f(t) = \frac{(r,-t)_P}{(r_1,-t)_P+(r_2,-t)_P}$$

Now since T_{r_1} and T_{r_2} have weighting measures and T_r is assumed to exist it must be given by

$$T_r = C\int_G U^t f(t) d(\overline{T}_{r_1}+\overline{T}_{r_2})(t) = C\int_G U^t d\overline{T}_r(t)$$

where

$$\overline{T}_r(A) = C\int_A f(t) d(\overline{T}_{r_1}+\overline{T}_{r_2})(t)$$

exists since f(t) is a bounded continuous real valued function and $\overline{T}_{r_1}$ and $\overline{T}_{r_2}$ are in M(G,C,R). r is therefore contained in the domain of definition of the analytic Laplace transform thereby verifying that J is an interval. Moreover, we may write T_r as

$$T_r = C\int_G U^t f(t) d(\overline{T}_{r_1}+\overline{T}_{r_2})(t)$$

$$= C\int_G \left[\frac{U^t(r,-t)_P}{(r_1,-t)_P+(r_2,-t)_P}\right] d(\overline{T}_{r_1}+\overline{T}_{r_2})(t)$$

whence the same argument as was used to compute the Fourier transform of an operator with a weighting measure will yield

$$\hat{T}(z)=\hat{T}_r(\gamma)=C\int_G\left[\frac{(\gamma,-t)_T(r,-t)_P}{(r_1,-t)_P+(r_2,-t)_P}\right]d(\overline{T}_{r1}+\overline{T}_{r2})(t)$$

$$=C\int_G\left[\frac{(z,-t)_{C_*}}{(r_1,-t)_P+(r_2,-t)_P}\right]d(\overline{T}_{r1}+\overline{T}_{r2})(t)$$

in the region $G_T x(r_1,r_2)$. Finally, since z only appears in the character $(z,-t)_{C_*}$ in this formula the analyticity of the character implies that of $\hat{T}(z)$ in the region where the above equality is valid (A.E.3), thereby completing the proof of the theorem.

Since the analytic Laplace transform coincides with the Laplace transform in its domain of definition, all of the properties stated for the latter apply to the former without modification. Of course, the sum and product of analytic functions is analytic hence the various operations preserve analyticity.

5. Causal Transforms

Although the Fourier transform contains all information about a subordinative operator, the causality structure of the operator is not apparent from the properties of the Fourier transform. This being primarily due to the fact that G_T is not ordered. On the other hand, the ordering of G_P allows one to deduce causality information from the Laplace transformation of an operator (SA-7). This is achieved with the aid of the following lemma.

Lemma: Let T be causal and f·T be well defined. Then f·T is causal.

Proof: Since E^s commutes with "dE(t)" and the scalar f(t) for all t if T is causal we have

$$E^S(f\cdot T) = E^S[C\int_G dE(t)[C\int_G f(t-q)TdE(q)]]$$

$$= C\int_G dE(t)[C\int_G f(t-q)E^S TdE(q)]$$

$$= C\int_G dE(t)[C\int_G f(t-q)E^S TE^S dE(q)]$$

$$= E^S[C\int_G dE(t)[C\int_G f(t-q)TdE(q)]]E^S = E^S(f\cdot T)E^S$$

which verifies the contention.

Our theorem on the transform of causal operators is:

<u>Theorem</u>: Let T be a bounded linear operator with analytic Laplace transform, $\hat{T}$, defined in a non-void region $G_T xJ$. Then T is causal if and only if J is unbounded on the right (i.e., if r_o is in J then r is in J whenever $r \geq r_o$) and $\hat{T}(z)$ is uniformly bounded on every set of the form $G_T x[r_o,\infty)$ with r_o in J. Moreover, when T is causal $\hat{T}$ is given by

$$\hat{T}(z) = C\int_G (z,-t)_{C_*}(-r_o,-t)_P d\overline{T}_{r_o}(t)$$

in every region of the form $G_T x[r_o,\infty)$.

Proof: If T is causal and its analytic Laplace transform exists at r_o then for any $r > r_o$

$$T_r=(r,-t)_P.T=(r-r_o,-t)_P(r_o,-t)_P.T=(r-r_o,-t)_P.T_{r_o}$$

hence

$$T_r=C\int_G U^t(r-r_o,-t)_P d\overline{T}_{r_o}(t)=C\int_G U^t d\overline{T}_r(t)$$

where the integral exists since $(r-r_o,-t)$ is a bounded continuous real valued function on the support of $\overline{T}_{r_o}$ (which is contained in $[0,\infty)$ since the causality of T implies that of T_{r_o} via the lemma) and

$$\overline{T}_r(A) = C\int_A (r-r_o,-t)_P d\overline{T}_{r_o}(t)$$

hence the analytic Laplace transform is defined at r. Moreover, upon invoking the isometric property of the Fourier transform we obtain via an argument similar to that used for computing the Fourier-Stieljes transform the formula

$$\hat{T}(z) = \hat{T}_r(\gamma) = C\int_G (\gamma,-t)_T (r-r_o,-t)_P d\overline{T}_{r_o}(t)$$

$$= C\int_G (z,-t)_{C_*} (-r_o,-t)_P d\overline{T}_{r_o}(t)$$

valid in $G_T x[r_o,\infty)$. Finally, since $(\gamma,-t)_T(r-r_o,-t)_P$ is bounded by 1 on the support of $\overline{T}_{r_o}$, $\hat{T}(z)$ is bounded by $|\overline{T}_{r_o}|_C$ in this region.

Conversely, if the analytic Laplace transform of T exists in a region which is unbounded and is a uniformly bounded function on every region of the form $G_T x[r_o,\infty)$ then it must be given by the above formula and we have for each r_o a real M such that

$$M \geq ||\hat{T}(z)|| = ||C\int_G (z,-t)_{C_*} (-r_o,-t)_P d\overline{T}_{r_o}(t)||$$

$$\geq ||C\int_{(-\infty,0)} (z,-t)_{C_*} (-r_o,-t) d\overline{T}_{r_o}(t)||$$

for all z in $G_T x[r_o,\infty)$. Now as r goes to infinity the function

$$(z,-t)_{C_*} (-r_o,-t)_P = (\gamma,-t)_T (r,-t)_P (-r_o,-t)_P$$

becomes unbounded in $(-\infty,0)$ and hence the above inequality cannot be satisfied unless the measure $\overline{T}_{r_o}$ is zero on $(-\infty,0)$. As such, $\overline{T}_{r_o}$ has its support contained in $[0,\infty)$ showing that T_{r_o} is causal (4.C.4) and hence by the lemma that

$$T = (-r_o,-t)_P \cdot T_{r_o}$$

is causal and completing the proof of the theorem.

One application of the preceeding result is to yield a condition for a causal operator to admit a causal inverse if one also requires that both the operator and its inverse admit analytic Laplace transforms.

Theorem: Let T be a causal bounded linear operator with analytic Laplace transform in a non-void region. Then a necessary and sufficient condition for T to have a causal bounded inverse which also admits an analytic Laplace transform in a non-void region is that for each r_o in J (where $G_T x J$ is the region of existence of the analytic Laplace transform of T where J is an interval which is unbounded on the right) there exist an $\varepsilon > 0$ such that

$$||T(z)|| \geq \varepsilon$$

for all z in $G_T x[r_o,\infty)$.

E. Problems and Discussion

1. Time-Invariant Operators

In addition to the (rather large number of) classes of time-invariant operators which we have already defined a number of interesting related classes may also be defined. For instance, we may term the uniform closure of operators

$$\sum_{i=1}^{n} R_i \hat{E}(A_i)$$

where R_i is static and A_i is a Borel set of G_T the Borel subordinative operators (KG-1, MS-1). These clearly contain the subordinatives which correspond to the special case when the A_i are open and hence may more accurately be termed the open subordinative operators and clearly they are time-invariant since they are the closure of a set of time-invariant operators.

Problem: Give examples of operators which are

i) time-invariant but not Borel subordinative.

ii) Borel subordinative but not (open) subordinative.

iii) open subordinative but do not admit a weighting measure representation.

Similarly, we may define classes of "character" subordinative operators to be the uniform closure of the operators

$$\sum_{i=1}^{n} R_i E(A_i)$$

where the A_i are taken either to be open sets of G or Borel sets of G. Such operators are clearly memoryless and are situated between the memoryless and static operators in a manner dual to the situation of the subordinative operators between the static and time-invariant operators. Clearly, most of our results on subordinative operators yield analagous results for these operators, the exceptions occuring only where the specific properties of ordered G's and their character groups have been invoked.

Problem: Formulate a "character" Fourier analysis for "character" subordinative operators via the equality

$$T = L\int_G \hat{T}(t)dE(t)$$

where $\hat{T}$ is a static operator valued function.

In general the properties of time-invariant operators are completely dual to those of memoryless operators; hence, our diagonal integral representation theory for memoryless operators (1.D.3) can be translated to the time-invariant case though we must work with a Lebesgue integral rather than a Cauchy integral since G_T is, in general, not ordered. We therefore denote by

$$L\int_{G_T} d\hat{E}(\gamma)Td\hat{E}(\gamma)$$

the uniform limit of the partial sums

$$\sum_{i=1}^{n} \hat{E}(A_i)T\hat{E}(A_i)$$

taken over the net of all Borel sets of G_T when it exists. Note that since we have not assumed that E has finite semi-variation with respect to the set of operators in which T lies this integral is not assured to exist as is the usual Lebesgue integral.

Problem: Show that an operator T is time-invariant if and only if

$$T = C\int_{G_T} d\hat{E}(\gamma)T d\hat{E}(\gamma)$$

Also upon following the results for the memoryless case we may term

$$L\int_{G_T} d\hat{E}(\gamma)T d\hat{E}(\gamma)$$

the time-invariant part of T when it exists.

Problem: Give a "physical" interpretation of the time-invariant part of an operator.

Problem: Determine the time-invariant part of the operator on $L_2(Z,R,m)$ defined by

$$y(k) = \sum_{j=-\infty}^{\infty} \underline{T}(k,j)u(j)$$

if it exists.

Since G_T is not ordered there is no natural analog of causality on the character space and hence upper and lower truncations of operators on $(H,\hat{E},\hat{U})$ are not defined. We do, however, conjecture that G_T does, in fact, admit some special characteristic different from, but as powerful as, ordering which would allow the defining of analogs of causal operators on the character space.

Although the various classes of time-invariant operators are symmetric with respect to time, the special representation applicable to time-invariant operators often simplify the characterization of causality. For instance, in the case of operators which admit a weighting measure representation.

Conjecture: A subordinative operator is causal if and only if it is the uniform limit of a sequence of causal trigonometric polynomials.

Since we have already characterized the causal trigonometric polynomials the validity of the above conjecture will allow a complete characterization of the causal subordinative operators and thus the extension of the results on the analytic Laplace transform of causal operators to the usual Laplace transform.

Also:

Conjecture: The inverse of a subordinative operator, if it exists, is subordinative.

2. Operator Decomposition

The various operator decomposition theorems formulated in our study of causality (1.F.1, 1.F.2, 1.F.3) can be strengthened if one restricts consideration to time-invariant operators.

Problem: Let an operator, T, be decomposed as

$$T = C + M + A$$

where

$$C = RC\int_G E_t T dE(t)$$
$$A = LC\int_G E^t T dE(t)$$

and

$$M = C\int_G dE(t) T dE(t)$$

and show that C, A, and M are time-invariant if and only if T is time-invariant.

Similarly, we can obtain a time-invariant multiplicative decomposition of a time-invariant operator via strengthening the lemma used to prove the general multiplicative decomposition (1.F.3).

Problem: Let T be a time-invariant contraction on a uniform

resolution space (H,E,U) and show that there exists a uniform resolution space $(\underline{H},\underline{E},\underline{U})$ and a time-invariant isometric extension Σ of T defined on $(H,E,U)\oplus(\underline{H},\underline{E},\underline{U})$ with matrix representation

$$\Sigma = \left[\begin{array}{c|c} T & 0 \\ \hline \Sigma_{21} & \Sigma_{22} \end{array}\right]$$

where Σ_{21} is a causal time-invariant operator mapping (H,E,U) to $(\underline{H},\underline{E},\underline{U})$ and Σ_{22} is a causal time-invariant operator mapping $(\underline{H},\underline{E},\underline{U})$ to $(\underline{H},\underline{E},\underline{U})$.

Problem: Let T be a bounded positive hermitian time-invariant operator on a uniform resolution space (H,E,U) and show that there exists a uniform resolution space $(\underline{H},\underline{E},\underline{U})$ and a causal time-invariant operator, C, mapping (H,E,U) into $(\underline{H},\underline{E},\underline{U})$ such that

$$T = C^*C$$

Note that unlike the general case wherein one has essentially no information about the extention space in the time-invariant case the representation theory for uniform resolution spaces (D.B.1) assures that $(\underline{H},\underline{E},\underline{U})$ is essentially the same as (H,E,U) since any two uniform resolution spaces over the same group differ at most by the cardinal number $e(\mu)$ (SA-6).

Not only are we assured that the decompositions will be time-invariant if T is time-invariant but, moreover, the existence of the additive decomposition is assured if T has a weighting function representation (SA-7).

Problem: Show that the operators, C, M and A in the above problem exist if T has a weighting measure representation. What are their weighting measures?

In fact, we believe that the above result can be somewhat strengthened via:

Conjecture: The operators C, M and A exist if T has an analytic Laplace transform in a non-void region.

Unfortunately, the above concepts cannot readily be extended beyond that of the above conjecture.

Problem: Give an example of a subordinative operator, T, for which the operators C, M and A do not all exist.

3. Semi-Uniform Resolution Space

A natural generalization of the concept of a uniform resolution space is a semi-uniform resolution space (H,E,V) where (H,E) is a resolution space and V is a semi-group or isometric operators defined in G^+ (the positive semi-group of G) satisfying

$$V^t E(A) = E(A+t)V^t$$

for all Borel sets A and t in G^+ (SA-8). Clearly, every uniform resolution space is semi-uniform. Possibly the most common class of semi-uniform space which is not uniform are the truncated uniform resolution spaces (H^+,E^+,U^+) where

$$H^+ = \{x \text{ in } H;\ E^0 x = 0\}$$

and E^+ and U^+ are the restrictions of E and U to H^+ where (H,E,U) is any specified uniform resolution space.

Problem: Show that every truncated uniform resolution space is a semi-uniform resolution space.

We denote the truncation of $L_2(G,K,m)$ by $L_2(G,K,m^+)$ which includes the usual L_2^+ and l_2^+ spaces, together with their unilateral shifts, when G = R or Z, respectively.

Problem: Show that for every truncated uniform resolution space that

$$(U^+)^t(U^+)^{t*} = E_t = 1 - E^t$$

for t in G^+.

Unlike the uniform case in a semi-uniform resolution space we say that an operator T is time-invariant if

$$V^tT = V^tV^{t*}TV^t$$

Problem: Show that the semi-uniform space and uniform space definitions for time-invariant coincide for operators on a uniform resolution space.

Problem: Show that an operator on a truncated uniform resolution space is time-invariant if and only if

$$(U^+)^tT = E_tT(U^+)^t$$

for all t in G^+.

Consistent with the results of the preceeding problems, the two definitions for time-invariance coincide on a uniform resolution space. That is, however, not the case for semi-uniform resolution space (SA-8, LA-1). In fact:

Problem: Show that an operator T on a truncated uniform resolution space is causal and time-invariant if and only if

$$(U^+)^tT = T(U^+)^t$$

for all t in G^+.

The above result allows one to solve the causal invertibility problem (1.C.3) for time-invariant operator on a truncated uniform resolution space via:

<u>Problem</u>: Show that if a causal time-invariant operator on a truncated uniform resolution space is invertible then the inverse is also causal and time-invariant.

Since the truncated uniform resolution space is a subspace of a uniform resolution space one would expect that most of the theory developed for operators on a uniform resolution space can be modified to cover operators on its truncation. Indeed, this is the case the modification being carried out via the following result (SA-8).

<u>Problem</u>: Let (H^+,E^+,U^+) be the truncation of the uniform resolution space (H,E,U) defined over a σ-compact group G. Then any causal time-invariant operator T^+ on (H^+,E^+,U^+) has a unique causal time-invariant extension T on (H,E,U).

Consistent with the above the theory developed for operators on a uniform resolution space also yields a theory for operators on a truncated uniform space, and, in fact, via the following result we also obtain a theory for operators on a semi-uniform resolution space.

<u>Problem</u>: Every semi-uniform resolution space with either $G = R$ or $G = Z$ is equivalent to the product of a uniform resolution space and a truncated uniform resolution space.

<u>Problem</u>: Show that every causal time-invariant operator on a semi-uniform resolution space with either $G = R$ or $G = Z$ can be uniquely extended to a causal time-invariant operator on a uniform resolution space.

Consistent with the above results virtually all of the properties of operators on a semi-uniform resolution space can be deduced from those of operators on a uniform resolution space without difficulty and hence there is little to gain by specifically studying operators on a semi-uniform space in the development of our theory.

4. Time-Invariant Feedback Systems

Although the Fourier and Laplace transforms yield a natural representation for time-invariant feedback systems which serves to tie together our abstract theory with the classical theory of feedback systems, time-invariance actually yields little in the way of strengthening the theory. The main area wherein stronger theorems are available is the study of the stability of feedback systems for which we have:

Problem: Let a feedback system defined on (H,E) have time-invariant open loop gain KF and show that it is stable if and only if the operator

$$I = E^{t_o}(KF) - E_{t_o}(KF)^*$$

has a bounded left inverse for t_o in G.

We note that our basic stability condition (2.B.3) amounts to a requirement that an operator have a causal inverse. This in turn was show to be equivalent to the condition that a family of operators have uniformly bounded left inverses (2.B.6). The above condition, however, requires that one only test the left invertibility of a single operator to determine the causal invertibility of the given time-invariant return difference. The price for this simplification, however, is that we must test the invertibility of a time variable operator rather than a time-invariant one.

Another manifestation of time-invariance in the stability problems results from the use of the Laplace transformation from which we obtain the "classical" Hurwitz test. This being essentially a statement of our Laplace transform condition for the existence of a causal bounded inverse for a causal bounded operator which has analytic Laplace transform in a non-void region (4.D.5).

Problem: Give a precise statement of the Hurwitz stability criterion for a time-invariant feedback system.

5. Time-Invariant Dynamical Systems

Given an operator T on a uniform resolution space (H,E,U) a state decomposition (3.A.1) (S,G(t),H(t)) for T is said to be a time-invariant state decomposition if

$$G(t)U^{-s} = G(t+s)$$

and

$$U^{s}H(t) = H(t+s)$$

for all t and s in G.

Problem: Show that (S,G(t),H(t)) is a time-invariant state decomposition if and only if there exists operators G and H mapping H to S and S to H respectively such that

$$G(t) = GU^{-t}$$

$$H(t) = U^{t}H$$

Note that in the above result H = H(0) and G = G(0) hence $H = E_0H$ and $G = GE^0$ and we usually denote a time-invariant state decomposition by (S,G,H) rather than $(S,GU^{-t},U^{t}H)$.

Problem: Show that every time-invariant operator admits a minimal time-invariant state decomposition.

Problem: Show that every time-invariant operator is regular.

Problem: Show that a strongly strictly causal operator is time-invariant if and only if it admits a time-invariant state decomposition.

The transition operators associated with a time-invariant state decomposition also exhibit a specialized form.

Problem: Let (S,G,H) be a time-invariant state decomposition with transition operators $\phi(t,q)$. Then show that for every $t \geq q$ and s in G

$$\phi(t,q) = \phi(t+s,q+s) = \phi(t-q,0)$$

Consistent with the above $\phi(t,q)$ is really only a function of one variable and by abuse of notation we write

$$\phi(t,q) = \phi(t-q) = \phi(v)$$

to emphasize that it is really only a function of $v = t - q$.

Problem: Show that if T is a linear operator with time-invariant state decomposition (S,G,H) and transition operators $\phi(v)$ then for any input u and $t \geq s$

$$x(t) = SRC\int_{(s,t)} \phi(t-q)GU^{-q}dE(q)u+\phi(t-s)x(s)$$

$$= SRC\int_{(-\infty,t)} \phi(t-q)GU^{-q}dE(q)u$$

The special form of G(t) and H(t) in a time-invariant state decomposition also allows a considerable simplification in our controllability and observability conditions.

Problem: Show that a time-invariant state decomposition (S,G,H) is controllable if and only if GG^* is uniformly positive definite and observable if and only if H^*H is uniformly positive definite.

Problem: Show that a time-invariant state decomposition is uniformly controllable if and only if

$$GE(-\delta,0)G^*$$

is uniformly positive definite for some $\delta > 0$ in G and uniformly observable if and only if

$$H^*E(0,\delta)H$$

is uniformly positive definite for some $\delta>0$, in G.

Note that the four controllability and observability operators used above all map S to itself hence in the common case (ZD-3) where S is finite dimensional these operators have simple matrix representations for which uniform positive definiteness reduces to positive definiteness which is readily tested.

6. Positive Transforms

If one characterizes the Laplace transform of the various operator classes defined by energy constraints the resolution space manifestation of such operator theoretic concepts as positive and bounded reality result, though without the reality condition if we allow complex Hilbert spaces (SA-10).

Problem: Let an operator T have a weighting measure and show that it is S-passive if and only if

i) Its analytic Laplace transform is defined on $G_T \times [0,\infty)$.

ii) $I-\hat{T}(z)^*\hat{T}(z)\geq 0$ for z in $G_T x[0,\infty)$.

Note that condition ii) above is equivalent to

$$||T(z)|| \leq 1$$

for z in $G_T x[0,\infty)$.

Problem: Let an operator T have a weighting measure and show that it is S-lossless if and only if

i) Its analytic Laplace transform is defined on $G_T x[0,\infty)$.

ii) $I-\hat{T}(z)^*\hat{T}(z)\geq 0$ for z in $G_T x[0,\infty)$.

iii) $I-\hat{T}(\gamma)^*\hat{T}(\gamma)=0$ for γ in G_T.

Problem: Let an operator T have a weighting measure and show that it is I-passive if and only if

i) Its analytic Laplace transform is defined on $G_T x[0,\infty)$.

ii) $\hat{T}(z)+\hat{T}(z)^*\geq 0$ for z in $G_T x[0,\infty)$.

Problem: Give a condition analogous to the above for I-losslessness.

Finally, it is also possible to characterize stable operators via the Laplace transformation.

Problem: Let an operator T have a weighting measure and show that it is stable if and only if:

i) Its analytic Laplace transform is defined on $G_T x[0,\infty)$.

ii) There exists a real constant M such that

$$||T(z)|| \leq M$$

for all z in $G_T x[0,\infty)$.

A. TOPOLOGICAL GROUPS

In many areas of applied mathematics the topological group proves to be a natural setting for a space of parameters. In our theory these parameters take the form of time and/or frequency, the relationship between the two being achieved via a Fourier or Laplace transformation, the topological group being the natural abstract setting for such transformations.

In the present chapter we define the basic group concepts which are needed for our theory. These include the basic definitions for the group concepts, a study of character groups and their duality theory, ordered groups and their representations, and the theory of integration and differentiation of functions defined on groups. Since the results are standard in Harmonic Analysis (RU-1, RE-1, HU-1, NA-1) no proofs are included.

In essence the material of the appendix may serve as a foundation for a study of Harmonic Analysis on Groups (RU-1) with the basic tools of that subject being formulated. Specific results of a Harmonic Analysis nature are, however, deferred to the appendix on spectral theory and the Fourier and Laplace transformations of Chapter 4.

A. Elementary Group Concepts

1. Definition and Examples

By an (LCA) group we mean an abelian group G with a locally compact Hausdorff topology in which the operation of translation $t \to t + s$ is continuous for all s and t in G. The most common examples of (LCA) groups are the additive groups of real numbers, R, integers, Z, and real or complex n-tuples, R^n or C^n; and the multiplicative groups of positive real numbers, P, non-zero complex numbers, C_*, and complex n-tuples of modulus 1, T^n. In general, we write all groups additively unless they have an explicit multiplicative structure.

2. Group homomorphisms

If G and H are (LCA) groups then a group homomorphism from G to H which is continuous in the topologies of the two groups is termed an (LCA) group homomorphism. All of the usual categorical concepts apply to the (LCA) groups using these morphisms (HU-1). In particular, a sub-group of G together with the relative topology is a sub-group in the category and the cartesian product of two groups together with the usual product topology is a product group in the category and similarly for the quotient group. Of course, two groups are equivalent if there exists a continuous group isomorphism between them. For instance, R and P are equivalent via $t \to e^t$ and C_* and PxT are equivalent via $a + ib \to (\sqrt{a^2+b^2}, \tan^{-1}(b/a))$, i.e., the polar decomposition.

B. Character Groups

1. Definition and Examples

We denote by M any multiplicative group of complex numbers such as T, P and C_* and term a group homomorphism defined on an (LCA) group G with values in M a character. In particular, for the cases of M=T, M=P and $M=C_*$ we term these group homomorphisms ordinary, real or complex characters, respectively. The set of all characters on G with values in M is denoted by G_M and the value of a character γ in G_M on an element t in G is denoted by $(\gamma,t)_M$. The set of ordinary characters with M=T has special properties, primarily since T is compact, and therefore plays a special role in much of our theory. As such we often term G_T the dual group for G.

The characters in G_M are given a group structure via $(\gamma+\delta.t)_M = (\gamma,t)_M(\delta.t)_M$. Clearly, we have:

$$(0,t)_M = (\gamma,0)_M = 1$$

and

$$(-\gamma,t)M = \overline{(\gamma,t)_M} = (\gamma,t)_M^{-1}$$

for all t in G. We give the character groups the compact-open topology via the neighborhood basis composed of sets of the form

$$U(K,\varepsilon) = \{\gamma \text{ in } G_M;\ |(\gamma,t) -1| < \varepsilon,\ t \in K\}$$

and it translates for each $\varepsilon > 0$ and compact set $K \subset G$. Clearly, the sets $U(K,\varepsilon)$ generate a well-defined Hausdorff topology (HU-1).

To see that the translation operation is continuous in the character group topology we let $\gamma_i \to \gamma$ in this topology hence it is contained in every set of the form $U(K,\varepsilon) + \gamma$ (a neighborhood basis for γ). Thus $\gamma_i + \delta$ is contained in every set of the form $U(K,\varepsilon) + \gamma + \delta$ and since this is a neighborhood basis for $\gamma + \delta$ we are assured that the sequence $\gamma_i + \delta$ converges to $\gamma + \delta$ in the topology of G_M showing that the translation operation is continuous.

For the real group R and any real γ the function $e^{-i\gamma t}$ is in R_T and in fact all members of R_T may be so obtained; hence R_T is isomorphic to R. Similarly, for the group Z, of integers, each real γ defines a character in Z_T via $e^{-i\gamma t}$ but since t must now be an integer the characters defined by γ and $\gamma + n2\pi$ coincide, hence only the real numbers between 0 and 2π define (distinct) ordinary characters and upon identifying this real interval with T we have Z_T isomorphic to T (RE-1). Similarly, T_T is isomorphic to Z. Of course, P is isomorphic to R hence P_T may be identified with R_T which in turn may be identified with P. By similar arguments we obtain $R_P \simeq R$, $Z_P \simeq P$ and $P_P \simeq P$. Of course, since $P \simeq R$ we may also identify R_P with P or Z_P with R, etc. Finally, upon invoking the fact that the isomorphism between C_* and TxP induces an isomorphism between G_{C_*} and $G_T x G_P$ we obtain $R_{C_*} \simeq RxR \simeq C$ and $Z_{C_*} \simeq TxP \simeq C_*$.

2. Products of Character Groups

The character groups associated with many common (LCA) groups can be obtained from the above examples via the following theorem (PO-1).

Theorem: For any two (LCA) groups, G and H, and any multiplicative group of complex numbers, M,

$$(GxH)_M \simeq G_M x H_M$$

Upon applying the preceeding result we have

$$C_{*T} \simeq (T\times P)_T \simeq T_T \times P_T \simeq Z\times R$$

Similarly, $(R^n)_T \simeq R^n$ and $(Z^n)_T \simeq T^n$, the n-torus formed by the complex numbers of modulus 1 in C^n. Of course, the same type of arguments also yield $(R^n)_P \simeq R^n$, $(Z^n)_P \simeq P^n$, $(R^n)_{C_*} \simeq R^{2n}$, and $(Z^n)_{C_*} \simeq {C_*}^n$.

3. <u>The Duality Theorem</u>

The dual group G_T has a number of special properties which are not shared by the other character groups. In particular, unlike G_P and G_{C_*} the dual group is always locally compact, hence an (LCA) group and, moreover, the dual of the dual group is isomorphic to the original group. This follows from the classical theorem of Pontryagin (PO-1 , RU-1) and is stated without proof as follows.

<u>Theorem</u>: For any (LCA) group G_T is an (LCA) group and moreover $(G_T)_T$ is isomorphic to G.

A number of powerful results can be obtained from the above duality theorem. In particular, the dual of every discrete group is compact and vice-versa (HU-1).

C. Ordered Groups

1. Positive Semigroups

An (LCA) group G is said to be ordered if there is given a positive semigroup $G^+ \subset G$ such that $G^+ \cap -G^+ = \{0\}$ and $G^+ \cup -G^+ = G$. G^+ induces a linear ordering on G via $t \geq s$ if $t - s$ is in G^+. We say that G is orderable if there exists a G^+ satisfying the above conditions. Of course, G^+ is not unique and, in fact, $-G^+$ also satisfies the required condition if G^+ satisfies them. We say that an (LCA) group has an archimedean ordering if it is ordered and for every pair of non-zero elements t and s in G^+ there exists an integer n such than $nt > s$.

2. A Representation Theorem

The class of orderable groups are characterized by the following theorem (RU-1).

Theorem: The following are equivalent for any (LCA) group G.

i) G is orderable.

ii) G is a discrete group containing no element of finite order or G is the product of R and a discrete group containing no element of finite order.

iii) G_T is a connected compact group or G_T is the product of R and a connected compact group.

Note that the discrete group in ii) may be the zero group; hence, R, itself, is orderable.

In the special case of an archimedean ordered group the above representation theorem can be further strengthened to:

Theorem: Let G be an archimedean ordered group. Then one (and only one) of the following is true

i) $G = R$

ii) G is a discrete subgroup of R.

3. Examples and Counter-Examples

The classical example of an ordered group is R with R^+ taken to be the non-negative real numbers though, of course, we could also take R^+ to be the non-positive real numbers which would reverse the linear ordering on R. Similarly, we may order Z by taking Z^+ to be the non-negative integers, and we may give Z^n a dictionary ordering (RE-1). In general, however, R^n $n \geq 2$ has no ordering of the required type, nor do T and C_*; these facts following from the representation theorem. On the other hand, the theorem assures that P is orderable since it is isomorphic to R.

Unlike the general (LCA) groups the orderable groups may not be closed under the usual categorical operations. In particular, the product of two orderable groups may not be orderable. In fact, $R^2 = R^1 \times R^1$ is not orderable. A subgroup of an orderable group is, however, orderable for given $H \subset G$ one may let $H^+ = H \cap G^+$.

In general the fact that G may be orderable does not imply that G_M is orderable and even when G_M may be orderable there may be no natural ordering. For instance $Z_T = T$ is not orderable whereas $R_T = R$ is orderable, but has no naturally induced ordering. On the other hand, G_P is always orderable if G is orderable via

$$G_P{}^+ = \{\gamma \in G_P \ ; \ (\gamma,t) \geq 1, \ t \in G^+\}$$

and hence one can define "strips" in G_{C_*} via the above ordering in G_P and the isomorphism identifying G_{C_*} and $G_T \times G_P$.

D. Integration on (LCA) Groups

1. Integration on Locally Compact Spaces

Since an (LCA) group is a locally compact topological space the standard theory for integrating scaler valued functions relative to a regular Borel measure defined on G is applicable (NA-1). We denote the set of all finite Borel measures on G by M(G) and order M(G) via absolute continuity (HA-1, NA-1), i.e., $\mu << \nu$ if μ is absolutely continuous with respect to ν. Similarly, we say μ is equivalent to ν if $\mu << \nu$ and $\nu << \mu$. Note that although M(G) explicitly contains only finite measures any σ-finite measure on a σ-compact group is equivalent to a finite measure, hence from the point of view of the equivalence classes M(G) contains all σ-finite regular Borel measures on G. Finally, if one takes advantage of the fact that G is a group we may define a translation operation in M(G) via

$$\mu_t(A) = \mu(A+t)$$

for all Borel sets A and t in G. Clearly if μ is in M(G) so are its translates and the translation operation preserves ordering and equivalence.

2. The Haar Measure

Unlike an arbitrary locally compact space the algebraic structure of an (LCA) group allows us to define special measures which are "well behaved" with respect to the group properties of G. Most important of these is the <u>Haar measure</u> which is defined to be a positive regular Borel measure satisfying the translation invariance condition

$$m(A) = m(A+t)$$

for all Borel sets A and t in G. i.e., m is a Haar measure if and only if

$m_t = m$ for all t in G.

On the groups of real or complex numbers the Lebesgue measure is a Haar measure as is the n dimensional Lebesgue measure on R^n or C^n since the translation of a cube does not change its Lebesgue measure. Similarly, the counting measure is a Haar measure for a discrete group since the translates of a set containing k elements has k elements. Of course, the measure induced on P via the Lebesgue measure on R under an isomorphism is a Haar measure as is the measure induced on T by the Lebesgue measure under its identification with a real interval.

The primary theorem concerning the existence and uniqueness of Haar measures is:

Theorem: Every (LCA) group admits a non-zero positive Haar measure which is unique up to a positive multiplicative constant.

Several proofs of this fundamental theorem for Harmonic analysis appear in (NA-1) and will not be given here. Since the non-trivial Haar measures on a group are unique except for a scale factor we generally refer to the Haar measure rather than a Haar measure.

3. Quasi-invariant Measures

We can weaken the translation invariant condition on the Haar measure by requiring only that $\mu(A+t) = 0$ if $\mu(A) = 0$. That is, the translates of sets of measure zero are also sets of measure zero. The resultant quasi-invariant measures satisfy many of the properties of the Haar measure and are often easier to deal with.

The quasi-invariant measures are characterized by the following theorem (MA-1).

Theorem: For a positive regular Borel measure μ on a σ-compact G the following are equivalent.

i) μ is quasi-invariant.

ii) μ is equivalent to m.

iii) μ is equivalent to μ_t for all t in $\mathfrak{G}$.

E. Differentiation on (LCA) Groups

1. Directional Derivatives

We would like to be able to define a directional derivative of a (Banach space valued) function defined on an (LCA) group. In general, however, the elements of the group cannot be identified with vectors. Hence, rather than defining a derivative in the direction of an element of the group, we define a derivative in the direction of a 1-parameter group; that is, an (LCA) group homomorphism from R into G. The set of such groups, denoted by W(G), form a real vector space via the operations

$$(u+v)(t) = u(t)+v(t)$$

and

$$(au)(t) = u(at)$$

for u and v in W(G) and real a. Thus we may talk of a directional derivative in the direction of a vector in W(G) rather than in the direction of an element in G.

If F is a (Banach space valued) function defined on an (LCA) group G and u is a 1-parameter group in W(G) we define the derivative of F at z in the u direction to be

$$F_u(z) = \lim_{t \searrow 0} \frac{F(z+u(t)) - F(z)}{t}$$

if it exists. In general, one may take the above limit in any of the standard Banach space topologies.

For the groups R^n and C^n the 1-parameter groups are the homomorphisms from R to R^n (R to C^n) which may be identified with the n by 1 matrices of real (complex) numbers and thus with R^n (C^n) itself. As such, in the case of $G = R^n$ or $G = C^n$ our vector space of directions coincides with the classical directions and hence our directional derivative reduces to the classical directional derivative.

As in the case of the classical derivative we have the following fundamental theorem (MA-2).

<u>Theorem</u>: Let $F_u(z)$ exist for all u in W(G) for some fixed z. Then $F_u(z)$ defines a linear operator mapping of W(G) into the Banach space in which F takes its values.

2. A Complex Structure for $W(C_*)$

Note that even though $W(C^n) \approx C^n$ we have only given W(G) a real vector space structure hence the linearity theorem says nothing about the homogeneity of $T_u(z)$ with respect to complex scalars. Although no complex vector space structure exists for W(G) in general, for instance if G = R, in the special case where G is the group of complex characters associated with some group we can give $W(G_{C_*})$ a complex vector space structure. Since both C^n and C_*^n are of this form such an approach allows one to treat the space of 1-parameter groups as a complex vector space in these classical cases.

The key to the formulation of the complex structure for $W(G_{C_*})$ is the following representation theorem (MA-2). Here a group homomorphism from G to R is termed a <u>linear functional</u> on G.

<u>Theorem</u>: Every 1-parameter group in $W(G_{C_*})$ can be written in the form

$$u(t) = e^{(u_1 + iu_2)t}$$

where u_1 and u_2 are linear functionals on G.

Note that the representation theorem essentially defines an isomorphism between $W(G_{C_*})$ and two copies of the vector space of all linear functionals on G.

Using the representation of the theorem we may now define the <u>scalar product</u> of a 1-parameter group in $W(G_{C_*})$ and a complex scalar via

$$(a+ib)e^{(u_1+iu_2)t} = e^{(au_1-bu_2+ibu_1+iau_2)t}$$

Although our definition makes use of several isomorphisms a little algebra will reveal that this is the usual scalar multiplication definition for those cases where $W(G_{C_*})$ is identifiable with a complex spcae. Of course, if $b = 0$ it reduces to the real scalar multiplication we have already defined.

3. <u>Analyticity</u>

In general, our theorem on the linearity of the derivative as a function from $W(G_{C_*})$ into a Banach space is false when complex scalars are employed. Of course, the additivity and real homogeneity of $F_u(z)$ remain valid in the complex case but $F_u(z)$ may not be homogeneous with respect to complex scalars, such as in the case of the complex conjugate. In the special case where $F_u(z)$ exists for all a and is homogeneous for all complex scalars we say that F is <u>analytic</u> at z. Of course, this coincides with the classical analyticity condition when $G_{C_*} = C^n$, the Cauchy-Riemann conditions serving as a test of whether or not $F_{iu}(z) = iF_u(z)$.

Standard arguments assure that constants are analytic as are the complex characters themselves (viewed as mappings from G_{C_*} to C_* with t in G fixed). Of course, as in the classical cases the analytic functions defined on G_{C_*} form a vector space.

B. OPERATOR VALUED INTEGRATION

Unlike scalar valued integrals wherein the Lebesgue integral yields an essentially universal theory of integration with the various alternative formulations yielding equivalent integrals, there are many alternative approaches to vector valued integration each possessing properties which are desirable for some applications and undesirable for others. In the present appendix we review those theories for the integration of operator valued functions over operator valued measures which are required for the main body of the work. Unlike classical spectral theoretic integrals (HA-1) which are equivalent to a family of scalar valued integrals, those integrals which are well defined for both operator valued functions and operator valued measures cannot be reduced to an equivalent scalar problem and thus must be formulated directly in a vector valued setting.

We begin with a survey of operator valued measures and the formulation of a generalized semivariation concept for operator valued measures with respect to operator valued functions. Since this constitutes a straightforward generalization of the classical semivariation theory (DI-1, DN-1) for integrating scalar or vector valued functions over operator valued measures the results on such measures and the corresponding Lebesgue integration theory are stated without proof. Possibly more important than the Lebesgue integral in our study is the Cauchy integral. Unlike the scalar valued case where the Cauchy and Lebesgue integrals coincide, in the operator valued setting the Cauchy integral has properties which are quite different than those of the Lebesgue integral (SA-1) and, in fact, the operator valued Cauchy integral can be defined with a number of alternative biases and/or the limits may be taken in various topologies. Moreover, in the main body of the text the correct choice of bias and/or topology for the Cauchy integrals proves to be of fundamental importance in our theory. As such, we study the properties of a number of alternative Cauchy integrals. Finally, we consider the special case of integration over spectral measures which arises in a number of distinct contexts in our work.

A. Operator Valued Measures

1. Definition and Examples

Throughout this appendix H is a Hilbert space, $\mathcal{B}$ is the (Banach) space of bounded linear operators on H, G is an (LCA) group and $\mathcal{C}$ and $\mathcal{D}$ are two subspaces of $\mathcal{B}$. Using this notation we term a $\mathcal{D}$ valued countably additive set function defined on the Borel sets of G an operator valued measure.

For our purposes the most important operator valued measure is the spectral measure, E, wherein E(A) is a hermitian projection for every Borel set A of G such that E(G) = I and $E(A) \leq E(\underline{A})$ (i.e. the range of E(A) is contained in the range of $E(\underline{A})$) if $\underline{A} \supset A$. Similarly, a generalized spectral measure corresponds to the case where P(A) is a positive hermitian operator for every Borel set of G such that P(G) = I and $P(A) \leq P(\underline{A})$ (i.e. $P(\underline{A}) - P(A)$ is a positive operator) if $\underline{A} \supset A$. Clearly, every spectral measure is a generalized spectral measure but not conversely (AK-1). Finally, we can define a class of "Stieltjes like" measures when G is an ordered group given an appropriate operator valued function, $M: G \to \mathcal{B}$, via the equality

$$M((-\infty,t)) = M(t)$$

for Borel sets of the form $A = (-\infty,t)$ with M being extended to the remainder of the Borel field in the natural manner (DI-1). In the special case where M is a strongly continuous non-decreasing projection valued function going strongly to the identity operator as t goes to ∞ and going strongly to the zero operator as t goes to $-\infty$ we term M a resolution of the identity and it induces a spectral measure via the above procedure. Finally, any scalar valued measure defines an operator valued measure by the equality

$$M(A) = \mu(A)I$$

for all Borel sets A. In general all of our operator valued integration theory applies to the scalar valued case via the above identification; hence, we integrate scalar valued functions and measures in an operator valued setting via the above identification, though, in general, we do not explicitly include the identity operator in our notation.

2. Semivariation

Given an operator valued measure, M, taking values in $\mathcal{D}$ we define its semivariation with respect to $\mathcal{C}$, $|M|_{\mathcal{C}}$, via

$$|M|_{\mathcal{C}} = \sup \left|\left| \sum_{i=1}^{n} T_i M(A_i) \right|\right|$$

where the supremum is taken over all finite partitions $\{A_i\}$ i=1, ..., n of G and all operators in $\mathcal{C}$ of norm less than or equal to 1.

For the case of the trivial measure μI induced by a scaler measure the semivariation is $\mu(G)$ independent of $\mathcal{C}$. On the other hand, non-trivial measures may have semivariation which is highly dependent on $\mathcal{C}$. For instance, if one lets $\mathcal{C}$ be the family of operators commuting with a spectral measure, E, we have for any partition, A_i, and family of operators, T_i, and u in H

$$\left|\left| \sum_{i=1}^{n} T_i E(A_i)u \right|\right|^2 = \left|\left| \sum_{i=1}^{n} E(A_i)T_i u \right|\right|^2 = \sum_{i=1}^{n} ||E(A_i)T_i u||^2$$

$$= \sum_{i=1}^{n} ||T_i E(A_i)u||^2 \leq \sum_{i=1}^{n} ||T_i||^2 ||E(A_i)u||^2$$

$$\leq \sum_{i=1}^{n} ||E(A_i)u||^2 = ||\sum_{i=1}^{n} E(A_i)u||^2 = ||u||^2$$

hence upon taking the appropriate supremums we have $|E|_{\mathcal{C}} = 1$ (since equality is obtained for $T_i = I$ for all i). On the other hand, if we let $\mathcal{C} = \mathcal{B}$ the summations in the definition of semivariation are unbounded and we have $|E|_{\mathcal{B}} = \infty$. Clearly, if $\underline{\mathcal{C}} \subset \mathcal{C}$ then $|M|_{\underline{\mathcal{C}}} \leq |M|_{\mathcal{C}}$; hence if $\mathcal{C}$ is taken to be the scalar operators (aI where a is a scalar) then $|E|_{\mathcal{C}} \leq 1$ and since the bound is clearly achieved we also have equality in this case.

3. $\underline{M(G,\mathcal{C},\mathcal{D})}$

We denote by $\underline{M(G,\mathcal{C},\mathcal{D})}$ the set of $\mathcal{D}$ valued regular measures with finite semivariation relative to $\mathcal{C}$. This notation is consistent with our previous notation for the scalar valued measures where we have $M(G) = M(G,\mathcal{C},\mathcal{D})$ for any space of operators $\mathcal{C}$ when $\mathcal{D}$ is the space of scalar operators. $M(G,\mathcal{C},\mathcal{D})$ is characterized by the following theorem (DI-1).

Theorem: $M(G,\mathcal{C},\mathcal{D})$ is a vector space normed by the semivariation.

B. The Lebesgue Integral

1. Totally Measurable Functions

Let M be a $\mathcal{D}$ valued measure having finite semivariation with respect to $\mathcal{C}$, then we say that a $\mathcal{C}$ valued function defined on G is totally measurable with respect to M if it can be uniformly approximated by $\mathcal{C}$ valued step function except possibly on a set of M-measure zero. Here the step functions are defined in the Borel sets of G in the usual manner.

We denote the set of totally measurable $\mathcal{C}$ valued function defined on G by $L_\infty(G,\mathcal{C},M)$ and for each f in $L_\infty(G,\mathcal{C},M)$ we define its norm by

$$||f||_\infty = \text{ess-sup}||f(t)||$$

Clearly, $L_\infty(G,\mathcal{C},M)$ is an algebra and by construction it is the completion in $||\ ||_\infty$ of a vector space of step functions hence:

Theorem: $L_\infty(G,\mathcal{C},M)$ normed by $||\ ||_\infty$ is a Banach Algebra.

2. Lebesgue Integration

The definition of the Lebesgue integral of a function in $L_\infty(G,\mathcal{C},M)$ is predicated on the following fundamental theorem (DI-1).

Theorem: Let f be in $L_\infty(G,\mathcal{C},M)$ and

$$s^i = \sum_{j=1}^{n(i)} s^i_j \chi_{A^i_k}$$

be a sequence of step functionin $L_\infty(G,\mathcal{C},M)$ uniformly approximating f (except possibly for a set of M-measure zero). Then

$$\lim_{i \to \infty} \sum_{j=1}^{n(i)} s^i_j M(A^i_j)$$

exists (in the uniform operator topology) and is independent of the choice of approximating sequence, s^i.

Consistent with the preceding theorem given any f in $L_\infty(G,\mathcal{C},M)$ we may define its <u>Lebesgue Integral</u> to be the common value of

$$\lim_{i \to \infty} \sum_{j=1}^{n(i)} s_j^i M(A_j^i)$$

where the s^i are any approximating sequence of step functions in $L_\infty(G,\mathcal{C},M)$ for f. We denote this limit by

$$L\int_G f(t)dM(t)$$

Clearly, the integral of a step function $s = \sum_{j=1}^{n} s_j \chi_{A_j}$ is $\sum_{j=1}^{n} s_j m(A_j)$ and hence the integral of a constant function equal to C for all t in G is CM(G).

The fundamental properties of the Lebesgue integral are characterized by the following theorem (DI-1).

<u>Theorem</u>: The Lebesgue integral is a linear continuous operator from $L_\infty(G,\mathcal{C},M)$ into $\mathcal{B}$ with norm equal to the semivariation of M with respect to $\mathcal{C}$.

In fact, the homogeneity of the theorem can be somewhat strengthened for if C is any operator in $\mathcal{C}$ then

$$L\int_G Cf(t)dM(t) = C[L\int_G f(t)dM(t)]$$

We also note that if f is in $L_\infty(G,\mathcal{C},M)$ then so is $f\chi_A$ for any Borel set, A, of G hence the Lebesgue integral of f over A is well defined via

$$L\int_A f(t)dM(t) = L\int_G f(t)\chi_A(t)dM(t)$$

3. <u>Riemann Integration</u>

Finally, we note that if f takes its value in a (uniformly) compact subset of $\mathcal{C}$ and is Borel measurable (i.e., $f^{-1}(B)$ is a Borel set for every ball, U, in $\mathcal{C}$) then it is in $L_\infty(G,\mathcal{C},M)$ (via the total boundedness of the compact set) and hence is Lebesgue integrable. In particular, bounded scalar (operator) valued functions are Lebesgue integrable.

One final subclass of "nice" functions which are always integrable are those members of $L_\infty(G,\mathcal{C},M)$ which are approximatable in the $\| \ \|_\infty$ norm, except possibly for a set of M measure zero, by $\mathcal{C}$ valued step functions of the form

$$\sum_{i=1}^{n} C_i \chi_{A_i}$$

where A_i is open for all i and C_i is in $\mathcal{C}$. By analogy to the scalar valued case these are termed the <u>Riemann totally integrable functions</u> and are denoted by $R_\infty(G,\mathcal{C},M)$. Clearly,

<u>Theorem</u>: $R_\infty(G,\mathcal{C},M)$ normed by $\| \ \|_\infty$ is a closed sub-algebra of $L_\infty(G,\mathcal{C},M)$ containing all totally measurable functions which are continuous M-almost everywhere.

4. <u>Integration with Measure on the Left</u>

Unlike the various scalar valued integrals wherein the value of the measure and the function commute in the operator valued case they may not commute hence we would like to define an alternative Lebesgue integral

$$L\int_G dM(t)f(t)$$

with the measure taken on the left of the function value in the partial sums used to define the integral of a totally measurable function in $L_\infty(G,\mathcal{C},M)$. Clearly, the theory for such integrals is precisely the same as for the standard Lebesgue integral and, in fact, one type of Lebesgue integral can be converted to the other by taking adjoints.

C. The Cauchy Integrals

1. Cauchy Integration

A major difficulty with the Lebesgue integral is that it is well defined only when M has finite semivariation with respect to $\mathcal{C}$. As such, the integral of many "simple" functions with respect to certain measures is undefined. For instance the integral of a $\mathcal{B}$ valued function relative to a spectral measure. This difficulty may, however, be circumvented when G is an ordered group by the Cauchy integrals. These, however, are defined without consideration of the topology on the space of functions being integrated and thus define unbounded linear operators with their inherent difficulties.

Let $\mathcal{C}$ and $\mathcal{D}$ be arbitrary and M by any $\mathcal{D}$ valued measure (possibly with infinite semivariation relative to $\mathcal{C}$) defined on an ordered group, G. Then the (uniform) Left Cauchy Integral of a function $f:G \to \mathcal{C}$ is defined by the equality

$$LC\int_G f(t)dM(t) = \underset{p \, \varepsilon \, P}{\text{limit}} \sum_{i=1}^{n(p)} f(t_{i-1})M(t_{i-1},t_i)$$

when the limit exists in the uniform topology over the net of all partitions of G into intervals $(t_{i-1}, t_i]$. Similarly, the (uniform) Right Cauchy Integral is defined by the equality

$$RC\int_G f(t)dM(t) = \underset{p \, \varepsilon \, P}{\text{limit}} \sum_{i=1}^{n(p)} f(t_i)M(t_{i-1},t_i)$$

When the limit exists in the uniform topology over the net of all partitions of G into intervals $(t_{i-1},t_i]$. Of course, we can also define Cauchy integrals with the measure on the left of the function in the obvious manner, or the Cauchy integral over a finite interval (s,t) for $s \leq t$

in G. In this latter case the integral is dependent only on s and t and not whether the interval of integration is open, closed, half-open, etc.

Unlike the Lebesgue integral which is assured to converge uniformly when f is in $L_\infty(G,\mathcal{C},M)$ the Cauchy integrable functions are defined by the convergence of the integral hence it is possible that the limit defining a Cauchy integral may converge in the strong operator topology when it fails to converge uniformly. In these cases we may define the strong Cauchy integrals, SLC$\int$ and SRC$\int$, with the measure on either side of f by taking the limit in the strong operator topology. Of course, we could also define weak Cauchy integrals though they do not play a role in the present development. In the remainder of the text unless specified to the contrary, all Cauchy integrals will be henceforth assumed to be uniformly convergent with the S preface always being used to denote the strong integrals. Of course, since uniform convergence implies strong convergence the existence of the uniform integrals implies that of the strong integrals but not conversely.

Finally, in cases where both the right and left (strong or uniform) Cauchy integrals exist and are equal (such as the scalar valued case relative to Lebesgue measure) and no "bias" is desired, we use the notation C$\int$ and SC$\int$.

2. Cauchy Integrable Functions

Although the space of Cauchy integrable functions is implicitly defined it is clearly linear (though not closed in general) and the integral defines a linear operator between this space and $\mathcal{B}$ (which is, in general, unbounded). Clearly, for a constant function equal to c for all t in G all of the Cauchy integrals exist and equal cM(G), and if (any of the)

Cauchy integrals of f exist then for any operator c in $\mathcal{C}$

$$c[C\int f(t)dM(t)] = C\int cf(t)dM(t)$$

where $C\int$ denotes an arbitrary Cauchy integral.

Unlike the Lebesgue integral for which the integrability of step function is immediate for the Cauchy integrals this is not the case. It is, however, possible to prove (SA-1).

Theorem: If s is any $\mathcal{B}$ valued step function defined on the semi-ring of left-closed right-open intervals, in G its (uniform, strong) Left Cauchy integral exists, and if s is any $\mathcal{B}$ valued step function defined on the semi-ring of left-open right-closed intervals, in G, its (uniform, strong) Right Cauchy integral exists.

Similarly, if f has a (uniform, strong) Left Cauchy integral over G then its (uniform, strong) Left Cauchy integral over a (possibly infinite) left-closed right-open interval, [a,b), defined by

$$LC\int_{[a,b)} f(t)dM(t) = LC\int_G f(t)\chi_{[a,b)}(t)dM(t)$$

exists and similarly, for the Right Cauchy integrals over sets of the form (a,b].

Although the Cauchy integrals relative to arbitrary functions and measures may have rather ugly properties if one makes a restriction on the measures and functions such as used to define the Lebesgue integral then the Cauchy integral is, itself, well behaved.

Theorem: Let M be in $M(G,\mathcal{C},\mathcal{D})$ and f be a continuous uniformly bounded complex valued function. Then the integrals

$$C\int_G f(t)dM(t)$$

$$RC\int_G f(t)dM(t)$$

$$LC\int_G f(t)dM(t)$$

and

$$L\int_G f(t)dM(t)$$

are all defined and equal.

3. A Counter-Example

Finally, we note that unlike the Cauchy integral of a scalar valued function relative to a scalar valued measure wherein the Left and Right Cauchy integrals always coincide (KE-1) in the operator valued case they may, indeed, be different. In fact, it is readily verified that if E is the spectral measure defined by a resolution of the identity E(t) then

$$LC\int_G E(t)dE(t) = 0 \neq I = RC\int_G E(t)dE(t)$$

In fact, one can construct examples wherein one Cauchy integral exists and the other does not exist (SA-1).

D. Integration over Spectral Measures

1. An Example

For many purposes the most interesting class of operator valued measure is the spectral measure. These are fundamental to much of classical spectral theory and also serve as a primary tool in much of modern system theory. The classical example of a spectral measure on the space $L_2(G,H,\mu)$ where H is a Hilbert space and μ is a σ-finite regular Borel measure on G, is the family of truncation operators defined by

$$(E(A)f)(s) = \begin{cases} f(s); & s \in A \\ 0 \;; & s \notin A \end{cases}$$

In fact, spectral multiplitity theory assures that every spectral measure is of this type, in an appropriately complex function space (HA-1). Finally, we note that the above operators, E(A), may also be interpreted as the operation of multiplying f by the characteristic function of A. That is;

$$(E(A)f)(s) = \chi_A(s)f(s)$$

2. The Idempotency Theorems

We have already seen that the semivariation of a spectral measure is one relative to the space of operators with which it commutes hence the Lebesgue integral of functions taking values in such a space (or a subspace thereof) is well defined over a spectral measure but not for more general valued functions. On the other hand, the Cauchy integrals over a spectral measure are meaningful for arbitrary operator valued functions though their existence is not assured a-priori, and as we have already seen, the Left and Right Cauchy integrals over a spectral measure may differ and/or be ill behaved.

The one special property attributable to both the Lebesgue and Cauchy integrals over a spectral measure is their multiplicative character which results from the idempotents property of the measure. That is: $E(A)E(A) = E(A)$ which leads to the following theorem (DL-1).

<u>Theorem</u>: Let $\int$ denote any of the operator valued integrals (Lebesgue, Cauchy, strong Cauchy, etc.) and let E be a spectral measure. Then

$$\int_G f(s)\int_G g(s,t)dE(t)dE(s) = \int_G f(r)g(r,r)dE(r)$$

and

$$\int_G dE(s)\int_G dE(t)g(s,t)f(s) = \int_G dE(r)g(r,r)f(r)$$

when all of the integrals involved exist.

Note that if the functions g and f take their values in a space of operators which commute with the spectral measure, E, we have as an immediate corollary that

$$\int_G f(s)dE(s)\int_G g(t)dE(t) = \int_G f(r)g(r)dE(r)$$

Another theorem which results from the idempotency of a spectral measure (DL-1) is:

<u>Theorem</u>: Let $\int$ denote any of the operator valued integrals (Lebesgue, Cauchy, strong Cauchy, etc.) and let E be a spectral measure. Then

$$\int_G f(t)dE(t)E(A) = \int_A f(t)dE(t)$$

and

$$E(A)\int_G dE(t)f(t) = \int_A dE(t)f(t)$$

for any Borel set A (interval for the Cauchy integrals) when all of the integrals involved exist.

C. SPECTRAL THEORY

Spectral theory, in its modern sense, may be viewed as the study of the representation of complex operators in terms of simple operators. The classical examples are, of course, matrix diagonalization and the representation of a normal operator as a Lebesgue integral of a complex valued function over a spectral measure. In the present appendix this latter theory manifests itself in the form of Stone's theorem wherein a group of unitary operators is represented by such an integral. When this is coupled with the representation theory for spectral measures as multiplication operators our desired simple representation for unitary groups is obtained. We then consider the problem of representing isometric and contractive semigroups via unitary groups whence a complete representation theory for the former is obtained. Although the results presented in this appendix are widely scattered in the literature (HA-1, FI-1, YS-1, SZ-1) they are, by now, classical and hence are presented without proof.

A. Spectral Theory for Unitary Groups

1. Unitary Groups

The purpose of spectral theory is to give a description (or representation) of an operator or family of operators in terms of simple operators which may be easily studied. With the ultimate goal of obtaining a representation for arbitrary semigroups of contractive operators we begin with the study of unitary groups of operators. That is; a strongly continuous function $U:G \to \mathcal{B}$, defined on an (LCA) group G whose values are unitary operators (in $\mathcal{B}$) on a Hilbert space such that

$$U^t(U^s)^{-1} = U^{t-s}$$

for all t and s in G. Clearly,

$$U^0 = U^{t-t} = U^t(U^t)^{-1} = 1$$

$$U^{-t} = U^{0-t} = 1(U^t)^{-1} = (U^t)^{-1}$$

and

$$U^tU^s = U^t(U^{-s})^{-1} = U^{t+s} = U^{s+t}$$

$$= U^s(U^{-t})^{-1} = U^sU^t$$

hence a unitary group is, indeed, well behaved.

Possibly the most common example of a group of unitary operators is the group of bilateral shifts, U, on the Hilbert space of L_2 functions (with values in any Hilbert space) relative to the Haar measure, $L_2(G,H,m)$, defined by the equality

$$(U^tf)(s) = f(s-t)$$

We note that this group is unitary because of the translation invariance of m (and if a more general measure were to be employed the operators of the shift group are not unitary and, in fact, may not be well defined (SA-2)). Alternatively, given any unitary operator on a Hilbert space we may define a unitary group over the integers, Z, via

$$U^n = \begin{cases} (U)^n; & n > 0 \\ 1 & ; n = 0 \\ (U^{-1})^{-n}; & n < 0 \end{cases}$$

In fact, every unitary group on Z is of this form where we take U to be U^1. Finally, for any spectral measure, E, defined on a group of ordinary characters, G_T, the Lebesgue integral defines a unitary group on G via

$$U^t = L\int_{G_T} (\gamma,-t)_T dE(\gamma)$$

Here the integral is assured to exist since a spectral measure has semi-variation one relative to scalar valued functions (B.A.2) and the group property results from the idempotent character of the spectral measure (B.D.2).

2. Stone's Theorem

The problem of representing a group of unitary operators (and hence a single unitary operator upon identifying it with U^1 in a unitary group over Z) can be converted to an equivalent (and hopefully easier) problem of representing a spectral measure by the following theorem (FI-1) of Stone which is essentially a converse to the preceeding example with the spectral measure.

Theorem: Every unitary group of operators, U, defined on a group

G is of the form

$$U^t = L\int_{G_T} (\gamma,-t)_T dE(\gamma)$$

where E is a spectral measure on the character group, G_T, uniquely determined by U.

In general the construction of the spectral measure associated with a unitary group via Stone's theorem is not a simple process. One special case where this can be achieved is the case of the shift group on $L_2(Z,R,m)$ wherein standard Fourier series techniques (YS-1) lead to the conclusion that U is represented by the spectral measure defined by multiplication by the characteristic function on $Z_T = T$.

Since there is a one-to-one correspondence between unitary groups and spectral measures any information contained in one is also contained in the other, though often in a completely different manner. One instance wherein this is not the case follows.

Theorem: Let U and T be a unitary group and spectral measure related via

$$U^t = L\int_{G_T} (\gamma,-t)_T dE(\gamma)$$

Then a bounded linear operator, A, commutes with U^t for all t in G if and only if it commutes with E(B) for all Borel sets B of G_T.

Stone's theorem yields an explicit formula for U in terms of E. On the other hand the construction of E, given U, is not explicitly characterized. Since E is a Borel measure it suffices to describe E(B) where B is open since the value of E on any other Borel set is uniquely determined by E(G) on the open sets. This is achieved via the following theorem.

Theorem: Let

$$p_n = \sum_{i=1}^{m(n)} c_i(\gamma, t_i)_T$$

be any sequence of trigonometric polynomials converging (uniformly almost everywhere) to χ_B, the characteristic function of an open set B (which exists via the Weierstrass approximation theorem). Then the sequence of operators

$$P_n = \sum_{i=1}^{m(n)} c_i U^{t_i}$$

converges uniformly to E(B).

B. Spectral Multiplicity Theory

1. Unitary Equivalence

Let U be a unitary group on G represented (via Stone's Theorem) by a spectral measure, E, on G_T and let V be an arbitrary unitary operator. Then

$$V^{-1}U^tV = L\int_{G_T}(\gamma,-t)_T dV^{-1}EV(\gamma)$$

where the integral is defined over the spectral measure whose value on a Borel set A is $V^{-1}E(A)V$. Conversely, this spectral measure defines a spectral measure which differs from U by the similarity transformation defined by V. As such the unitary equivalence classes of the group U are preserved by Stone's theorem and we may study these classes by studying the unitary equivalence classes for the spectral measures.

2. The Spectral Multiplicity Function

Our invariant for the spectral measures, and hence by the preceeding development for the unitary groups, is the spectral multiplicity function. This is defined as a function, e, mapping the set of finite Borel measures on G into a set of cardinal numbers satisfying;

i) $e(0) = 0$ (i.e. the multiplicity of the zero measure is zero).

ii) e is decreasing in the sense that if $0 \neq \nu \ll \mu$ then $e(\mu) \leq e(\nu)$ where "<<" denotes absolute continuity.

iii) If $\mu = \sum_i \mu^i$ where the μ^i are pairwise orthogonal non-zero measures then $e(\mu) = \min_i\{e(\mu^i)\}$.

Note that ii) assures that a spectral multiplicity function is an equivalence class invariant in M(G) and hence its domain may be extended to σ-finite Borel measures.

3. Canonical Representation

Although we have defined the spectral multiplicity function on all of M(G) it is entirely determined by its values on an appropriately chosen countable family of measures in M(G) termed a canonical representation of e. The measures, μ^i, making up such a representation must

i) be pairwise orthogonal ($\mu^i \perp \mu^j$; $i \neq j$)

ii) have uniform multiplicity (i.e., whenever $0 \neq \nu << \mu^i$ then $e(\nu) = e(\mu^i)$.)

iii) and satisfy the equality

$$e(\nu) = \begin{cases} \min e(\mu^i); \ \mu^i \not\perp \nu \\ 0 \ ; \text{ if } \mu^i \perp \nu \text{ for all } i \end{cases}$$

Here "$\perp$" denotes orthogonality and "$\not\perp$" denotes non-orthogonality of measures.

Although a canonical representation is required to satisfy a rather strong set of axioms at least one such representation is assured to exist for each spectral multiplicity function. In fact, it can be shown (HA-1):

Theorem: Every spectral multiplicity function has a unique (up to equivalence classes in M(G)) canonical representation satisfying the condition $e(\mu^i) \neq e(\mu^j)$ if $i \neq j$.

4. The Spectral Multiplicity Theorem

Finally, with the spectral multiplicity function and its canonical representation as a tool we may formulate our primary representation theorem for the spectral measures and hence via Stone's theorem also for unitary groups of operators. For this purpose we let E be a spectral measure on the Borel sets of a group, G, whose values are projections on a Hilbert space H over a field of scalars F (=R or C); we let $L_2(G, F, \mu)$ be the

Hilbert space of scaler valued functions defined on G which are square integrable relative to a σ-finite Borel measure, μ; and we let $L_2(G,F,\mu)^\alpha$, α a cardinal number, denote α copies of $L_2(G,F,\mu)$. We then have the following fundamental theorem (HA-1).

<u>Theorem</u>: To each spectral measure E one may uniquely associate a spectral multiplicity function, e, in such a manner that two spectral measures are unitarily equivalent if and only if they have the same spectral multiplicity function. Moreover, if μ^i is any canonical representation of e then E is unitarily equivalent to the spectral measure defined by multiplication by the characteristic function on the space

$$\sum_i \oplus L_2(G,F,\mu^i)^{e(\mu^i)}$$

In essence, the theorem not only yields a unitary invariant for the spectral measures but also a representation of any spectral measure in terms of the "simplest" possible such measure, though on an extremely complex function space.

Of course, upon combining the preceeding result with Stone's theorem we obtain a unitary invariant and representation for every unitary group wherein one associates a spectral multiplicity function with each unitary group in such a manner that two groups are unitarily equivalent if and only if they have the same spectral multiplicity function.

C. Spectral Theory for Contractive Semigroups

1. Contractive Semigroups

Let G be an (LCA) group ordered by a positive semigroup, G^+, and let T be a strongly continuous function defined on G^+ whose values are contractive operators in $\mathcal{B}$, satisfying

$$T^t T^s = T^{t+s}$$

Then we say that T is a semigroup of contractive operators. Clearly,

$$T^0 T^t = T^{0+t} = T^t$$

hence $T^0 = I$ and

$$T^t T^s = T^{t+s} = T^{s+t} = T^s T^t$$

Any unitary group of operators on G is a unitary semigroup when restricted to G^+. Another class of contractive semigroup which is similar to the above unitary semigroup is an isometric semigroup wherein V^t is an isometric operator for all t in G^+. An example of such a semigroup is the semigroup of unilateral shifts on $L_2(G,H,m^+)$ defined by

$$(V^t f)(s) = \begin{cases} f(s-t); & \text{if } s-t \in G^+ \\ 0 & ; \text{if } s-t \notin G^+ \end{cases}$$

for all s and t in G^+. Here m^+ is the Haar measure on G restricted to G^+ and H is any Hilbert space. Finally, given any contractive operator, T, we may define a contractive semigroup on Z^+ by letting

$$T^n = (T)^n$$

Clearly, any contractive semigroup on Z^+ is of this form if one lets $T=T^1$

and, as such, the theory of single contractive operators is essentially equivalent to the theory of contractive semigroups on Z^+.

2. Representation of Isometric Semigroups

Our approach to the representation of contractive semigroups is to convert their representation into an equivalent problem of representing unitary groups which we have already solved. The first step in this conversion results from the following theorem which allows one to decompose an isometric semigroup into a unitary group (restricted to G^+) and a unilateral shift group (FI-1).

Theorem: Let G be either the group of real numbers or the group of integers with G^+ taken to be the usual positive semigroup, and let H be a Hilbert space. Then for any isometric semigroup, V, defined on G^+ whose values are operators on H there exists Hilbert spaces K_1 and K_2 and a unitary transformation from H to $K_1 \oplus L_2(G,K_2,m^+)$ which takes V into the product of a unitary semigroup on K_1 and the unilateral shift semigroup on $L_2(G,K_2,m^+)$.

Although the theorem is valid more generally than G=R or G=Z, it is not valid for arbitrary ordered groups. Since we have already characterized the unitary groups (and hence also the unitary semigroups) while the unilateral shift group is readily described the theorem yields a complete characterization of the isometric semigroups.

3. Isometric Extension

Finally, we consider the problem of converting the study of a general contractive semigroup (on Z^+ or R^+) to the study of an isometric semigroup, this being achieved by extension of the domain of the operators T^t to a larger space on which the extended operator is isometric. For a contractive semigroup, T, defined on G^+ whose values are operators on a Hilbert space H we define its <u>isometric extension</u> to be an isometric semigroup, V, defined on G^+ whose values are operators on $H \oplus K$, K a Hilbert space, such that

$$T^t = P_H V^t \Big|_H \triangleq prV^t$$

If one represents the operators on $H \oplus K$ as a 2 by 2 matrix of operators then the isometric extension, V^t, or T^t is of the form

$$V^t = \left[\begin{array}{c|c} T^t & V^t_{12} \\ \hline V^t_{21} & V^t_{22} \end{array}\right]$$

We say that an isometric extension is <u>minimal</u> if K is spanned by the vectors of the form $\{V^t_{21}x \; ; \; x \text{ in } H, t \in G^+\}$.

Since prV^t is clearly contractive for any isometric V^t we certainly cannot hope to construct isometric extensions of any class of operators more general than the contractive semigroups. In fact, every such semigroup has an isometric extension (FI-1).

<u>Theorem</u>: Let G be either the group of real numbers or the group of integers with G^+ taken to be the usual positive semigroup, and let H be a Hilbert space. Then every contractive semigroup, T, defined on G^+ whose values are operators on H has a minimal isometric extension which is unique up to a unitary equivalence.

Although the construction of an isometric extension of a contractive semigroup defined on R^+ may be quite complex the process is quite straightforward for contractive semigroups defined on Z^+. Given such a T we define an operator, A, by the equality

$$A = (I - T^{1*}T^1)^{1/2}$$

where the square root is assured to exist (RI-1) since $I-T^{1*}T^1$ is positive and hermitian for a contractive T^1, and we let D be the range of A and K be the direct product of countably many copies of D. Now, any x in H⊕K is of the form $x = (h, d_1, d_2, d_3, \ldots)$ and we may define V^1x by the equality

$$V^1(h, d_1, d_2, d_3, \ldots) = (T^1h, Ah, d_1, d_2, d_3)$$

and $V^n = (V^1)^n$. Now the verification that V is the required isometric extension of T is immediate.

D. REPRESENTATION THEORY

Unlike a classical Hilbert space wherein any two spaces of the same dimension are equivalent, dimension thereby serving as a complete set of invariants for the category of Hilbert spaces, when dealing with resolution space we require that an equivalence transformation preserve the time properties of the space as well as its Hilbert space properties. As such, two resolution spaces of the same dimension may not be equivalent and the representation theory for the category of resolution spaces (with causal maps) is considerably more complex than for the category of Hilbert spaces. Fortunately, however, the time structure in a resolution space is defined by a spectral measure for which we already have a representation theory (C,B,4) which may be adopted to the resolution space problem. In fact, this theory yields a representation of every resolution space as a function's space together with the usual resolution structure, though in general the function space is quite complex. Moreover, in the case of a uniform resolution space the representation yields an L_2 space of Hilbert space valued functions defined with respect to Haar measure, the dimension of the Hilbert space in which the functions take their values yielding a complete set of invariants. As such, uniform resolution spaces, like Hilbert spaces themselves, are characterized by a simple dimensional parameter whereas a spectral multiplicity function is required for the characterization of general resolution space.

A. Resolution Space Representation Theory

1. Resolution Space Equivalence

We say that two resolution spaces, (H,E) and $(\underline{H},\underline{E})$ are equivalent if there exists a causal Hilbert space isomorphism (i.e. isometry) from H onto $\underline{H}$ which has a causal inverse. This is the standard definition for equivalence in the category of Hilbert spaces together with causal operators (FR-1); and, in fact, all of the categorical concepts which we have considered such as product and sub-resolution space reduce to the abstract categorical definitions for these concepts in this category. In essence, the equivalence concept requires that the Hilbert spaces H and $\underline{H}$ be equivalent in such a way as to preserve the time structure as well as the usual Hilbert space structure.

The equivalence transformations between a pair of resolution spaces are characterized by the following theorem:

Theorem: Let V be a mapping from a resolution space (H,E) to a resolution space $(\underline{H},\underline{E})$. Then the following are equivalent.

i) V is a resolution space equivalence.

ii) V is a memoryless Hilbert space isomorphism

iii) V is a congruence, between E and $\underline{E}$ (i.e., $E = V^{-1}\underline{E}V$).

Proof: i) => ii). If V is a resolution space equivalence transformation it is causal hence its adjoint is anti-causal, but since V is isometric its adjoint is equal to its inverse which is causal. V^{-1}(and thus also V) is therefore both causal and anti-causal hence memoryless verifying ii).

ii) => iii). If V is a memoryless isometry then

$$VE(A) = \underline{E}(A)V$$

for all Borel sets A hence

$$E(A) = V^{-1}\underline{E}(A)V$$

verifying iii).

iii) => i). Finally, if V is a congruence between E and $\underline{E}$ it is memoryless (via the above argument reversed) and hence is a causal, causally invertible isometry from H to $\underline{H}$, i.e., a resolution space equivalence. The proof of the theorem is thus complete.

2. A Representation Theorem

Consistent with the preceeding theorem the problem of characterizing equivalent resolution spaces and the problem of characterizing unitarily equivalent spectral measures are the same. As such, we can obtain a complete representation theory for the category of resolution spaces and causal maps by translating the standard theory for spectral measures from classical spectral theory (C.B.4).

Theorem: For each resolution space (H,E) there is a uniquely defined spectral multiplicity function, e, such that two resolution spaces are equivalent if and only if they have the same spectral multiplicity function. Moreover, for any canonical representation, μ^i of e, (H,E) is equivalent to

$$\sum_i \oplus L_2(G,F,\mu^i)^{e(\mu^i)}$$

with its usual spectral measure. Here F is the field over which H is defined and $L_2(G,F,\mu^i)^{e(\mu^i)}$ denotes $e(\mu^i)$ copies of $L_2(G,F,\mu^i)$.

Note that the above theorem implies that all resolution spaces can be viewed as L_2 function spaces with the "usual" time structure, though in general relative to a very complicated measure. The classical Hilbert space result that any two spaces of the same dimension are equivalent, however, fails when applied to resolution spaces since it may be impossible to find a Hilbert space isomorphism between two spaces of the same dimension that also preserves the time-structure (KP-1). Moreover, even when a space is equivalent to a function space with the usual spectral measure the transformation may be so complicated as to preclude use of this fact (DU-1, KI-1). For instance, the resolution spaces of stochastic processes whose spectral measure estimates the future on the basis of past data is relatively simple as defined but becomes extremely complex if one attempts to transform it into an equivalent function space with the usual spectral measure.

For the space $L_2(G,K,m)$, K a Hilbert space and m the Haar measure on a σ-compact G, with its usual resolution structure we may decompose m into a countable sum of pairwise orthogonal finite measures μ^i since m is σ-finite (HA-2) and equate $L_2(G,K,m)$ with

$$\sum_i \oplus L_2(G,F,\mu^i)^{\dim(K)}$$

As such the theorem assures that μ^i is canonical representation for e with $e(\mu^i) = \dim(K)$ for all i, hence e is characterized by

$$e(\mu) = \begin{cases} \dim(K) & ; \text{ if } m >> \mu \neq 0 \\ 0 & ; \text{ otherwise} \end{cases}$$

Clearly, the dimension of K is the only invariant for this special, but important, class of spaces.

B. Uniform Resolution Space Representation Theory

1. Uniformable Resolution Spaces

Upon invoking our representation theory for general resolution spaces as a tool we may now formulate a representation theory for uniform resolution spaces via a result which is somewhat reminiscent of the Mackey Imprimativity theorem (MA-1, MS-3). Before considering the properties of the shift operators themselves we characterize the class of resolution spaces (H,E) which admit a uniform structure i.e., a group of unitary operators satisfying the equality

$$U^t E(A) = E(A+t)U^t$$

We say that such a space is uniformable. If a family of operators; U^t, t in G; satisfy the above equality but do not satisfy the group condition we say that the resolution space is partially uniformable.

In the following theorem (H,E) is a resolution space with spectral multiplicity function e and we say that e is translation invariant if for all μ in M(G) and t in G

$$e(\mu_t) = e(\mu)$$

where μ_t is the translate of μ defined by

$$\mu_t(A) = \mu(A+t)$$

for all Borel sets A.

Theorem: For a resolution space (H,E) over a σ-compact group G with spectral multiplicity function e the following are equivalent.

i) (H,E) is partially uniformable.

ii) (H,E) is equivalent to (H,E_t), where E_t is the translate of E, for all t in G.

iii) e is translation invariant.

iv) (H,E) is equivalent to $L_2(G,K,m)$ with its usual resolution structure where K is a Hilbert space and m is the Haar measure on G.

v) (H,E) is uniformable.

Proof: i) => ii). If (H,E) is partially uniformable then for each t a unitary operator U^t satisfying

$$U^tE(A) = E(A+t)U^t$$

exists for every Borel set A, hence since $E(A+t) = E_t(A)$ we have the equality

$$U^tE(A) = E_t(A)U^t$$

showing that U^t is a memoryless transformation between (H,E) and (H,E_t) and since it is also unitary it is a resolution space equivalence varifying ii).

ii) => iii). Assume that ii) holds and let e be the spectral multiplicity function for (H,E) (C.B.4) with canonical representation μ^i and let e_t be the spectral multiplicity function for (H,E_t). Now since E_t differs from E only by a time shift μ_t^i is a canonical representation for e_t, hence upon invoking the resolution space representation theory

$$(H,E) \simeq \sum_i L_2(G,F,\mu^i)^{e(\mu^i)}$$

and

$$(H,E_t) \simeq \sum_i L_2(G,F,\mu_t^i)^{e_t(\mu_t^i)}$$

both with their usual resolution structure. Now, since the $e(\mu^i)$ are simply cardinal numbers independent of t, shifting the time parameter in our representation of (H,E) yields

$$(H,E_t) \simeq \sum_i L_2(G,F,\mu_t^i)^{e(\mu^i)}$$

Now by ii) $(H,E) \simeq (H,E_t)$ hence $e = e_t$ yielding

$$(H,E_t) \simeq \sum_i L_2(G,F,\mu_t^i)^{e(\mu_t^i)}$$

Finally, upon equating our last two representations for (H,E_t) we have

$$\sum_i L_2(G,F,\mu_t^i)^{e(\mu^i)} \simeq \sum_i L_2(G,F,\mu_t^i)^{e(\mu_t^i)}$$

but since the μ_t^i are pairwise orthogonal this can only hold if $e(\mu^i) = e(\mu_t^i)$ for all i. e is thus translation invariant on the μ^i but since they form a canonical representation e must be translation invariant on all of M(G) verifying iii).

iii) => iv). Assume that e is translation invariant and let μ^i be a canonical representation of e which we may, without loss of generality, assume satisfies the condition

$$e(\mu^i) \neq e(\mu^j)$$

if and only if $i \neq j$ (C.B.3). Now, we claim that each μ^j in this canonical representation is quasi-invariant (A.D.3). If this were not the case for some μ^k there would exist a Borel set A and t in G such that $\mu^k(A) \neq 0$ but $\mu^k(A+t) = 0$ hence we may define a measure μ_A^k in M(G) by

$$\mu_A^k(B) = \mu^k(B \cap A)$$

for all Borel sets B. Now clearly, $\mu_A^k << \mu^k$ hence it is orthogonal to each μ^i, $i \neq k$ but it is not orthogonal to μ^k (since A is a common set of non-zero measure), and since μ^i is a canonical representation for e it has uniform multiplicity hence

$$e(\mu_A^k) = e(\mu^k)$$

On the other hand, since e is assumed to be translation invariant

$$e(\mu_A^k) = e((\mu_A^k)_t) = \min\{e(\mu^i);\ (\mu_A^k)_t \not\perp \mu^i\} \neq e(\mu^k)$$

where the first equality is due to the translation invariance of e, the second to the fact that the μ^i form a canonical representation of e and the third equality results from the fact that $\mu^k(A+t) = 0$ implies that $(\mu_A^k)_t \perp \mu^k$ while, by assumption $e(\mu^i) \neq e(\mu^k)$ when $i \neq k$. Since our last two equations are contradictory our assumption that μ^k is not quasi-invariant must be false.

Since on a σ-compact group any quasi-invariant measure is equivalent to the Haar measure (C.B.3) the above contention implies that all of the μ^i in the canonical representation of e are equivalent to one and another (since they are all equivalent to m), but since μ^i is a canonical representation they are also pairwise orthogonal. Now these two contradictory requirements can be simultaneously satisfied if and only if there is but one measure in the canonical representation, say μ, which is quasi-invariant. A resolution space with translation invariant e is therefore assured to be equivalent to

$$L_2(G,F,\mu)^{e(\mu)}$$

where μ is quasi-invariant. Finally, since μ is quasi-invariant it is equivalent to the Haar measure, m, since G is σ-compact. Moreover, since G is σ-compact the Radon-Nykodym theorem (HA-2) assures that μ and m are related via

$$\mu(A) = \int_A f(t)dm(t)$$

where f is a measurable function which is strictly positive almost everywhere. As such we may define a transformation

$$T: L_2(G,F,\mu)^{e(\mu)} \to L_2(G,F,m)^{e(\mu)}$$

via $(Tx)(t) = f(t)^{1/2}x(t)$ for each function x in $L_2(G,F,\mu)^{e(\mu)}$. Now, T is clearly memoryless since multiplication by f commutes with multiplication by characteristic functions. Moreover,

$$\int_G ||(Tx)(t)||^2 d\mu(t) = \int_G ||f(t)^{1/2}x(t)||^2 d\mu(t)$$

$$= \int_G ||x(t)||^2 f(t)d\mu(t) = \int_G ||x(t)||^2 dm(t)$$

hence T is an isometric transformation between $L_2(G,F\mu)^{e(\mu)}$ and $L_2(G,F,m)^{e(\mu)}$. Finally, since f is strictly positive i.e., T^{-1} is the operator of multiplication by $1/f(t)^{1/2}$. T is therefore a resolution space equivalence between $L_2(G,F,\mu)^{e(\mu)}$ and $L_2(G,F,m)^{e(\mu)}$ completing our proof that the translation invariance of e implies that (H,E) is equivalent to $L_2(G,F,m)^{e(\mu)}$. Now, upon identifying $L_2(G,F,m)^{e(\mu)}$ with $L_2(G,K,m)$ where K is an $e(\mu)$ dimensional Hilbert space over F we have the desired result.

<u>iv) => v)</u>. Clearly, the usual shift operators are well defined on $L_2(G,K,m)$

hence iv) implies v).

v) => i). If a resolution space is uniformable it is also partially uniformable hence v) implies i) thereby completing the proof.

Consistent with the theorem a uniform resolution space on a given group is completely determined by the cardinal number $e(\mu)$, where μ is any quasi-invariant measure in M(G), hence we term this number the multiplicity of (H,E,U).

2. Shifts on $L_2(G,F,m)$

Consistent with the preceeding theorem all uniform resolution spaces over a σ-compact G are equivalent to direct products of $L_2(G,F,m)$; hence, the problem of characterizing all uniform resolution spaces reduces to the problem of characterizing all possible families of shift operators on this space with its usual resolution structure.

Theorem: Every unitary group of shift operators on $L_2(G,F,m)$ with its usual resolution structure is of the form

$$(U^t f)(s) = g(s,t)f(t-s)$$

where $g(s,t)$ is a function defined on GxG whose values are scalars of magnitude one satisfying

$$g(s,t-r) = g(s,t)g(s-t,-r)$$

Moreover, every such g defines a group of shift operators.

Proof: We note that the usual resolution structure on $L_2(G,F,m)$ is the operation of multiplication by the characteristic function hence we denote both the operator E(A) and the characteristic function of A by χ_A the correct interpretation being apparent from the context. Now if U is any group of shift operators we have

$$U^t\chi_A = U^t\chi_A\chi_A = \chi_{A+t}U^t\chi_A$$

where the rightmost χ_A in each term is a characteristic function and the others are operators of multiplication by the characteristic function. Now for each Borel set A, and s and t in G define the function g(A,s,t) by

$$g(A,s,t) = (U^t\chi_A)(s)$$

Now

$$g(A,s,t) = (U^t\chi_A)(s) = (\chi_{A+t}U^t\chi_A)(s) = \chi_{A+t}(s)(U^t\chi_A)(s)$$

which is zero if $s \notin A+t$ and its value is independent of A if $s \in A+t$. To see this let s be in A+t and s also be in B+t. Then

$$\begin{aligned} g(A,s,t) &= (U^t\chi_A)(s) = \chi_{B+t}(s)(U^t\chi_A)(s) \\ &= (\chi_{B+t}U^t\chi_A)(s) = (U^t\chi_B\chi_A)(s) = (U^t\chi_{A\cap B})(s) \\ &= (U^t\chi_A\chi_B) = (\chi_{A+t}U^t\chi_B)(s) = \chi_{A+t}(s)(U^t\chi_B)(s) \\ &= (U^t\chi_B)(s) = g(B,s,t). \end{aligned}$$

Consistent with the above we may therefore define g(s,t) to be the common value of g(A,s,t) for s in A+t. Clearly we have

$$g(A,s,t) = g(s,t)\chi_{A+t}(s) = (U^t\chi_A)(s)$$

Now any function in $L_2(G,F,m)$ is a limit of a sequence of linear combinations of characteristic functions hence by the linearity and continuity of U^t we have for any f in $L_2(G,F,m)$

$$f = \lim_{k \to \infty} \sum_{i=1}^{n(k)} c_i^k \chi_{A_i^k}$$

where the c_i^k are scalars and

$$(U^t f)(s) = U^t[\lim_{k \to \infty} \sum_{i=1}^{n(k)} c_i^k \chi_{A_i^k}]$$

$$= \lim_{k \to \infty}[\sum_{i=1}^{n(k)} c_i^k (U^t \chi_{A_i^k})](s)$$

$$= \lim_{k \to \infty}[\sum_{i=1}^{n(k)} c_i^k (g(s,t)\chi_{A_i^k + t}](s)$$

$$= g(s,t)[\lim_{k \to \infty} \sum_{i=1}^{n(k)} c_i^k \chi_{A_i^k + t}] = g(s,t)f(s-t)$$

as required. Finally, since U is a group we obtain the equality

$$g(s,t-r) = g(s,t)g(s-t,-r)$$

and since U^t are unitary we deduce the fact that $|g(s,t)| = 1$ for all t and s in the obvious way. Moreover, the fact that a g satisfying the above conditions defines a shift group follows from direct calculation. With the assumed verification of these facts the proof is thus complete.

3. Uniform Equivalence

The preceeding theory yields a complete characterization of the resolution space equivalence classes of uniformable resolution spaces. This theory, however, is predicated on the equivalence concepts for general resolution spaces whereas when dealing with uniform resolution space it is natural to define a stronger equivalence concept which preserves the shift properties of the space as well as its time properties. We therefore say that two uniform resolution spaces, (H,E,U) and ($\underline{H}$,$\underline{E}$,$\underline{U}$) are *uniformly* equivalent if there exists a causal, time-invariant Hilbert space equivalence mapping (H,E,U) onto ($\underline{H}$,$\underline{E}$,$\underline{U}$) which admits a causal time-invariant inverse. As in the resolution space equivalence concept such a transformation must be memoryless (D.A.4) hence a uniform equivalence transformation is a static (i.e., memoryless and time-invariant) isometric mapping from H onto $\underline{H}$, and clearly, uniform equivalence implies equivalence.

Theorem: Two uniform resolution spaces, (H,E,U) and ($\underline{H}$,$\underline{E}$,$\underline{U}$) are uniformly equivalent if and only if they are equivalent and for some equivalence transformation, K, the integral

$$T = C\int_{\underline{G}} \underline{U}^{-t} K U^{t} dE(t)$$

exists.

Proof: Since K is an equivalence it is memoryless hence for any Borel set A we have

$$\begin{aligned} TE(A) &= C\int_{\underline{G}} \underline{U}^{-t} K U^{t} dE(t)E(A) = C\int_{\underline{G}} \underline{U}^{-t} K U^{t} E(A) dE(T) \\ &= C\int_{\underline{G}} \underline{U}^{-t} K E(A+t) U^{t} dE(t) = C\int_{\underline{G}} \underline{U}^{-t} \underline{E}(A+t) K U^{t} dE(t) \\ &= \underline{E}(A)\ C\int_{\underline{G}} \underline{U}^{-t} K U^{t} dE(t) = \underline{E}(A)T \end{aligned}$$

hence T is memoryless when it exists and hence if the integral exists it may also be written in the form

$$T = C\int_{G} d\underline{E}(t)\underline{U}^{-t} K U^{t} = C\int_{G} d\underline{E}(t)\underline{U}^{-t} K U^{t} dE(t)$$

We therefore have

$$\begin{aligned} T^*T &= \left(C\int_{G} d\underline{E}(t)\underline{U}^{-t} K U^{t}\right)^* \left(C\int_{G} d\underline{E}(s)\underline{U}^{-s} K U^{s}\right) \\ &= \left(C\int_{G} U^{-t} K^* \underline{U}^{t} d\underline{E}(t)\right)\left(C\int_{G} dE(s)\underline{U}^{-s} K U^{s}\right) \end{aligned}$$

$$= C\int_G U^{-t}K^*\underline{U}^t d\underline{E}(t)\underline{U}^{-t}KU^t = C\int_G U^{-t}K^*\underline{U}^t\underline{U}^{-t}KU^t dE(t)$$

$$= C\int_G 1\ dE(t) = 1$$

and similarly

$$TT^* = 1$$

hence T is a Hilbert space equivalence. Finally, for any s in G we have

$$TU^s = C\int_G d\underline{E}(t)\underline{U}^{-t}KU^tU^s = C\int_G dE(t)\underline{U}^{-t}KU^{t+s}$$

$$= C\int_G d\underline{E}(r-s)\underline{U}^{s-r}KU^r = C\int_G d\underline{E}(r-s)\underline{U}^s\underline{U}^{-r}KU^r$$

$$= \underline{U}^s\ C\int_G d\underline{E}(r)\underline{U}^{-r}KU^r = \underline{U}^sT$$

Thus T is time-invariant and a uniform resolution space equivalence if the hypotheses of the Theorem are satisfied. Conversely, if T is a uniform equivalence then it is an equivalence and

$$T = \underline{U}^{-t}TU^t$$

for all t since it is time-invariant hence

$$C\int_G \underline{U}^{-t}TU^t dE(t) = C\int_G TdE(t) = T$$

exists and the hypotheses of the theorem are verified as required to complete the proof.

References

AK-1 Akhiezer, N.I. and I.M. Glazman, Theory of Linear Operators on Hilbert Space, New York, Ungar, 1963.

AN-1 Anderson, B.D.O., "An Algebraic Solution to the Spectral Factorization Problem," IEEE Trans., Vol. AC-12, pp. 410-414, (1967).

AN-2 Anderson, B.D.O., "A Distributional Approach to Time-Varying Sensitivity," SIAM Jour. on Appl. Math., Vol. 15, pp. 1001-1010 (1967).

BA-1 Baxter, G., "An Operator Identity," Pacific Jour. of Math., Vol. 8, pp. 649-663, (1963).

BE-1 Beltrami, E.J., "Dissipative Operators, Positive Real Resolvants and the Theory of Distributions," in Applications of Generalized Functions, Philadelphia, SIAM, 1966.

BL-1 Balakrishnan, A.V., "Linear Systems with Infinite Dimensional State Spaces," Proc. of the Conf. on System Theory, Poly. Inst. of Brooklyn, pp. 69-96, 1965.

BL-2 Balakrishnan, A.V., "On the State Space Theory for Linear Systems," Jour. of Math. Anal. and Appl., Vol. 14, pp. 371-391, (1966).

BL-3 Balakrishnan, A.V., "State Space Theory of Linear Time-Varying Systems," in System Theory (ed. L. Zadeh and E. Polak), New York, McGraw-Hill, pp. 95-126, 1969.

BL-4 Balakrishnan, A.V., "Foundations of the State Space Theory of Continuous Systems," Jour. of Comp. and Sys. Sci., Vol. 1, pp. 91-116, (1967).

BL-5 Balakrishnan, A.V., "On the State Space Theory of Nonlinear Systems," in Functional Analysis and Optimization (ed. E.R. Caianello), New York, Academic Press, pp. 15-36, 1966.

BR-1 Brodskii, M.S., Triangular and Jordan Representation of Linear Operators, Providence, AMS, 1971.

CH-1 Choquet, G., Lectures on Analysis, New York, W.A. Benjamin, Inc., 1969.

DA-1 Damborg, M.J., "Stability of the Basic Nonlinear Operator Feedback System," Rech. Rpt. 37, Systems Engrg. Lab., Univ. of Michican, 1967.

DA-2 Damborg, M.J., "The Use of Normed Linear Spaces for Feedback System Stability," Proc. of the 14th Midwest Symp. on Circuit Thy., Univ. of Denver, 1971.

DA-3 Damborg, M.J., and A. Naylor, "Fundamental Structure of Input-Output Stability for Feedback Systems," IEEE Trans., Vol. SSC-6, pp. 92-96, (1970).

DE-1 DeSantis, R.M., "Causality Structure of Engineering Systems," Ph.D. Thesis, Univ. of Michigan, 1971.

DE-2 DeSantis, R.M., "Causality and Stability in Resolution Space," Proc. of the 14th Midwest Symp. on Circuit Thy., Univ. of Denver, 1971.

DE-3 DeSantis, R.M., "Causality, Strict Causality and Invertibility for Systems in Hilbert Resolution Space," Unpublished Notes, Ecole Polytechnique, 1971.

DE-4 DeSantis, R.M., "On the Solution of the Linear Volterra Equation in Hilbert Space," Unpublished Notes, Ecole Polytechnique, 1971.

DE-5 DeSantis, R.M., "On A Generalized Volterra Equation in Hilbert Space," Proc. of the AMS, (to appear).

DE-6 DeSantis, R.M., and W.A. Porter, "On Time Related Properties of Nonlinear Systems," SIAM Jour. on Appl. Math. (to appear).

DE-7 DeSantis, R.M., "Causality for Nonlinear Systems in Hilbert Space," Unpublished Notes, Ecole Polytechnique, 1972.

DE-8 DeSantis, R.M., "On State Realization and Causality Decomposition for Nonlinear Systems," Proc. of the 41st ORSA Meeting, New Orleans, 1972.

DE-9 DeSantis, R.M., and W.A. Porter, "Temporal Properties of Engineering Systems," Proc. of the 9th Allerton Conf. on Circuit and Sys. Thy., Univ. of Illinois, 1971.

DE-10 DeSantis, R.M., and W.A. Porter, "Causality and Multilinear Operators," Unpublished Notes, Univ. of Michigan, 1970.

DI-1 Dinculeanu, N., Vector Measures, Oxford, Pergamon, 1967.

DL-1 Daleckii, J.R., and S.G. Krein, "Integration and Differentiation of Functions of Hermitian Operators and Their Applications to the Theory of Perturbations," AMS Translations, Ser. 2, Vol. 47, pp. 1-30, (1967).

DN-1 Dunford, N., and J.T. Schwartz, Linear Operators, New York, Interscience, 1958.

DO-1 Dolph, C., "Positive Real Resolvants and Linear Passive Hilbert Systems," Annales Acad. Sci. Fennicae, Ser. A, Vol. 336/9, (1963).

DS-1 Desoer, C.A. and M.Y. Wu, "Input-Output Properties of Multiple-Input Multiple-Output Discrete Systems," Jour. of the Franklin Inst., Vol. 290, pp. 11-24 and 85-101, (1970).

DU-1 Duttweiler, D.L., "Reproducing Kernel Hilbert Space Techniques for Detection and Estimation Problems," Tech. Rpt. 7050-18, Information Sci. Lab., Stanford Univ., 1970.

FA-1 Falb, P.L., and M.I. Freedman, "A Generalized Transform Theory for Causal Operators, SIAM Jour. on Cont., Vol. 7, pp. 452-471, (1969).

FA-2 Falb, P.L., "On a Theorem of Bochner," Publications Mathematiques No. 37, Institut Des Hautes Etudes Scientifiques, 1969.

FE-1 Freedman, M.I., Falb, P.L., and G. Zames, "A Hilbert Space Stability Theory over Locally Compact Abelian Groups," SIAM Jour. on Cont., Vol. 7, pp. 479-493, (1969).

FE-2 Freedman, M.I., Falb, P.L., and J. Anton, "A Note on Causality and Analyticity," SIAM Jour. on Cont., Vol. 7, pp. 472-478, (1969).

FI-1, Filmore, P., Notes on Operator Theory, New York, Van Nostrand-Reinhold, 1970.

FO-1 Foures, Y., and I.E. Segal, "Causality and Analyticity," Trans. of the AMS, Vol. 78, pp. 385-405, (1955).

FR-1 Freyd, P., Abelian Categories, New York, Harper and Row, 1964.

GE-1 Gersho, A., "Nonlinear Systems with a Restricted Additivity Property," IEEE Trans., Vol. CT-16, pp. 150-154, (1969).

GO-1 Gohberg, I.C., and M.G. Krein, "On Factorization of Operators in Hilbert Space," Societ Math. Dokl., Vol. 8, pp. 831-834, (1967).

GO-2 Gohberg, I.C., and M.G. Krein, Introduction to the Theory of Linear Nonselfadjoint Operators in Hilbert Space, Providence, AMS, 1967.

GO-3 Gohberg, I.C., and M.G. Krein, Theory of Volterra Operators In Hilbert Space and its Applications, Providence, AMS, 1970.

GO-4 Gohberg, I.C., and M.G. Krein, "Systems of Integral Equations on the Half Line with Kernels Depending on the Difference of Their Arguments," AMS Translations, Ser. 2, Vol. 14, pp. 217-287, (1960).

HA-1 Halmos, P.R., Introduction to Hilbert Space and the Theory of Spectral Multiplicity, New York, Chelsea, 1951.

HA-2 Halmos, P.R., Measure Theory, Princeton, Van Nostrand, 1948.

HU-1 Husain, T., Introduction to Topological Groups, Philadelphia, W.B. Saunders, 1966.

KA-1 Katznelson, Y., Harmonic Analysis, New York, J. Wiley and Sons, 1968.

KE-1 Kelly, J.L., General Topology, Princeton, Van Nostrand, 1955.

KG-1 Kolmogorov, A.N., "Stationary Sequences in Hilbert Space", Byull. Moskov. Gos. Univ. Math., Vol. 2, (1941).

KI-1 Kailath, T., and D.L. Duttweiler, "Generalized Innovation Processes and Some Applications," Proc. of the 14th Midwest Symp. on Circuit Thy., Univ. of Denver, 1971.

KL-1 Klir, G., An Approach to General Systems Theory, New York, Van Nostrand-Reinhold, 1969.

KM-1 Kalman, R.E., "Algebraic Structure of Linear Dynamical Systems," Proc. of the Nat. Acad. of Sci. (USA), Vol. 54, pp. 1503-1508, (1965).

KM-2 Kalman, R.E., "Algebraic Aspects of Dynamical Systems," in Differential Equations and Dynamical Systems (ed. J.K. Hale and J.P. LaSalle), New York, Academic Press, pp. 133-146, (1967).

KM-3 Kalman, R.E., "Canonical Structure of Linear Dynamical Systems," Proc. of the Nat. Acad. of Sci. (USA), Vol. 48, pp. 596-600, (1962).

KM-4 Kalman, R.E., Arbib, M.A., and P.L. Falb, Topics in Mathematical System Theory, New York, McGraw-Hill, 1969.

KO-1 Kou, M.C., and L. Kazda, "Minimum Energy Problems in Hilbert Space," Jour. of the Franklin Inst., Vol. 283, pp. 38-54, (1967).

KP-1 Kaplan, W., Ph.D. Thesis, Univ. of Maryland, 1970.

KR-1 Krein, M.G., "Integral Equations on the Half Line with Kernels Depending on the Difference of their Arguments, "AMS Translations, Ser. 2, Vol. 22, pp. 163-288, (1956).

LA-1 Leake, R.J., "Discrete Time Systems," Unpublished Notes, Univ. of Notre Dame, 1971.

LE-1 Levan, N., "Active Characteristic Operator Function and the Synthesis of Active Scattering Operators," Unpublished Notes, Univ. of California at Los Angeles, 1971.

LE-2 Levan, N., "Synthesis of Scattering Operator by its Characteristic Operator Function, "Unpublished Notes, Univ. of California at Los Angeles, 1971

LE-3 Levan, N., "Synthesis of Active Scattering Operator by its m-Derived Passive Operator," IEEE Trans., Vol. CT-19, pp. 524-526, (Sept. 1972).

LE-4 Levan, N., "Lossless Extension of Active Scattering Operators and Synthesis of these Operators," Proc. of the Inter. Symp. on Circuit Theory, Los Angeles, 1972.

LO-1 Loomis, L., An Introduction to Abstract Harmonic Analysis, Princeton, Van Nostrand, 1953.

MA-1 Mackey, G., "A Theorem of Stone and Von Neumann," Duke Math. Jour., Vol. 16, pp. 313-320, (1949).

MA-2 Mackey, G., "The Laplace Transform for Locally Compact Abelian Groups," Proc. of the Nat. Acad. of Sci. (USA), Vol. 34, pp. 156-162, (1948).

ME-1 Mesarovic, M., "Foundations of System Theory," in Views of General System Theory, (ed. M. Mesarovic), New York, J. Wiley and Sons, pp. 1-24, 1964.

MS-1 Masani, P., "The Normality of Time-Invariant Subordinative Operators in Hilbert Space," Bull. of the AMS, Vol. 71, pp. 546-550, (1965).

MS-2 Masani, P., "Orthogonally Scattered Measures," Adv. in Math., Vol. 2, pp. 61-117, (1968).

MS-3 Masani, P., "Quasi-Isometric Measures and their Applications," Bull of the AMS, Vol. 76, pp. 427-528, (1970).

MS-4 Masani, P., and M. Rosenberg, "When is an Operator the Integral of a Spectral Measure," Unpublished Notes, Univ. of Indiana, 1971.

MS-5 Masani, P., "Factorization of Operator Valued Functions," Proc. of the London Math. Soc., Vol. 6, Ser. 3, pp. 59-64, (1956).

NA-1 Nachbin, L., The Haar Integral, Princeton, Van Nostrand, 1965.

PO-1 Pontyagin, L., Topological Groups, Princeton, Princeton Univ. Press, 1946.

PR-1 Porter, W.A., "Some Circuit Theory Concepts Revisited," Int. Jour. on Cont., Vol. 12, pp. 433-448, (1970).

PR-2 Porter, W.A., "Nonlinear Systems in Hilbert Space," Int. Jour. on Cont., Col. 13, pp. 593-602, (1971).

PR-3, Porter, W.A., "A Basic Optimization Problem in Linear Systems," Math. Sys. Thy., Vol. 5, pp. 20-44 (1971).

PR-4 Porter, W.A., and C.L. Zahm, "Basic Concepts in System Theory," Tech. Rpt. 33, Systems Engineering Lab., Univ. of Michigan, 1969.

PR-5 Porter, W.A., Modern Foundations of System Theory, New York, MacMillan, 1966.

PR-6 Porter, W.A., "Some Recent Results in Nonlinear System Theory," Proc. of the 13th Midwest Symp. on Circuit Theory, Univ. of Minnesota, 1970.

RE-1 Reiter, H., Classical Harmonic Analysis and Locally Compact Groups, Oxfor Oxford Univ. Press, 1968.

RI-1 Riesz, F., and B. sz-Nagy, Functional Analysis, New York, Ungar, 1960.

RO-1 Rosenberg, M., "Time Invariant Subordinative Operators in A Hilbert Space," Notices of the AMS, Vol. 16, pp. 641, 1969.

RU-1 Rudin, W., Fourier Analysis on Groups, New York, Interscience, 1967.

SA-1 Saeks, R., and R.A. Goldstein, "Cauchy Integrals and Spectral Measures," Indiana Math. Jour., (to appear).

SA-2 Saeks, R., "On the Existence of Shift Operators," Tech. Memo. EE-707b, Univ. of Notre Dame, 1970.

SA-3 Saeks, R., "Resolution Space - A Function Analytic Setting for Control Theory," Proc. of the 9th Allerton Conf. on Circuit and System Theory, Univ. of Illinois, 1971.

SA-4 Saeks, R., "Causality in Hilbert Space," SIAM Review, Vol. 12, pp. 357-383, (1970).

SA-5 Saeks, R., "State in Hilbert Space," SIAM Review (to appear).

SA-6 Saeks, R., "Causal Factorization, Shift Operators and the Spectral Multiplicity Function," Tech. Memo. EE-715b, Univ. of Notre Dame, 1969.

SA-7 Saeks, R., "Fourier Analysis in Hilbert Space," SIAM Review, (to appear).

SA-8 Saeks, R., and R.J. Leake, "On Semi-Uniform Resolution Space," Proc. of the 14th Midwest Symp. on Circuit Theory, Univ. of Denver, 1971.

SA-9 Saeks, R., "Synthesis of General Linear Networks," SIAM Jour. on Appl. Math., Vol. 16, pp. 924-930, (1968).

SA-10 Saeks, R., Generalized Networks, New York, Holt, Rinehart and Winston, 1972.

SC-1 Schnure, W.K., "Controllability and Observability Criteria for Causal Operators on a Hilbert Resolution Space," Proc. of the 14th Midwest Symp. on Circuit Thy., Univ. of Denver, 1971.

SI-1 Sain, M.K., "On Control Applications of a Determinant Equality Related to Eigen-Value Computation," IEEE Trans., Vol. AC-11, pp. 109-111, (1966).

SN-1 Sandberg, I., "Conditions for the Causality of Nonlinear Operators Defined on a Linear Space," Quart. on Appl. Math., Vol. 23, pp. 87-91, (1965).

SN-2 Sandberg, I., "On the L_2 Boundedness of Solutions of Nonlinear Functional Equations," Bell Sys. Tech. Jour., Vol. 43, pp. 1601-1608, (1968).

SO-1 Solodov, A.V., Linear Automatic Control Systems with Varying Parameters, New York, Am. Elsevier, 1966.

SA-1 sz-Nagy, B., and C. Foias, Analyse Harmonique Des Operateurs de L'Espace De Hilbert, Paris, Masson, 1967.

WE-1 Wiener, N., "On the Factorization of Matrices," Comment Math. Helv., Vol. 29, pp. 97-111, (1955).

WI-1 Winslow, L., and R. Saeks, "Nonlinear Lossless Networks," IEEE Trans., Vol. CT-19, p. 392, (July 1972).

WL-1 Willems, J.C., "Stability, Instability, Invertibility and Causality," SIAM Jour. on Cont., Vol. 7, pp. 645-671, (1969).

WL-3 Willems, J.C., "The Generation of Lyapunov Functions for Input-Output Stable Systems," SIAM Jour. on Cont., Vol. 9, pp. 105-134, (1971)

WL-4 Willems, J.C., and R.W. Brockett, "Some New Rearrangement Inequalities" IEEE Trans., Vol. AC-13, pp. 539-549. (1968).

WL-5 Willems, J.C., Analysis of Feedback Systems, Cambridge, MIT Press, 1971

WN-1 Windeknecht, T.G., "Mathematical System Theory - Causality," Math. Sys. Thy., Vol. 1, pp. 279-288, (1967).

WN-2 Windeknecht, T.C., General Dynamical Processes, New York, Academic Press, 1971.

WO-1 Wohlers, M.R., and E.J. Beltrami, Distributions and Boundary Values of Analytic Functions, New York, Academic Press, 1966.

YO-1 Youla, D.C., Castriota, L.J., and H.J. Carlin, "Bounded Real Scattering Matrices and the Foundations of Linear Passive Network Theory," IRE Trans., Vol. CT-6, pp. 102-124, (1959).

YS-1 Yosida, K., Functional Analysis, New York, Springer-Verlag, 1966.

ZA-1 Zames, G., "On Input-Output Stability of Time Varying Nonlinear Feedback Systems, " IEEE Trans., Vol. AC-11, pp. 228-238, and 465-476, L966).

ZA-2 Zames, G., "Functional Analysis Applied to Nonlinear Feedback Systems," IEEE Trans., Vol. CT-10, pp. 392-404, (1963).

ZA-3, Zames, G., and P.L. Falb, "Stability Conditions for Systems with Monotone and Slope-Restricted Nonlinearities," SIAM Jour. on Cont., Vol 6, pp. 89-108, (1968).

ZD-1 Zadeh, L., "The Concept of State in System Theory," in Views on General System Theory, (ed. M. Mesarovic), New York, J. Wiley and Sons, pp. 39-50, 1964.

ZD-2 Zadeh, L., "The Concepts of System Aggregate and State in System Theory," in System Theory, (ed. L. Zadeh and E. Polak), New York, McGraw-Hill, pp. 3-42, 1969.

ZD-3 Zadeh, L., and C.A. Desoer, Linear System Theory, New York, McGraw-Hill, 1963.

ZE-1 Zemanian, A.H., "The Hilbert Port," SIAM Jour. on Appl. Math., Vol. 18, pp. 98-138, (1970).

ZE-2 Zemanian, A.H., "A Scattering Formalism for the Hilbert Port," SIAM Jour. on Appl. Math., Vol. 18, pp. 467-488, (1970).

ZE-3 Zemanian, A.H., "Banach Systems, Hilbert Ports and ∞-Ports," Unpublished Notes, State Univ. of New York at Stoney Brook, 1971.

Lecture Notes in Economics and Mathematical Systems

(Vol. 1–15: Lecture Notes in Operations Research and Mathematical Economics, Vol. 16–59: Lecture Notes in Operations Research and Mathematical Systems)

Vol. 1: H. Bühlmann, H. Loeffel, E. Nievergelt, Einführung in die Theorie und Praxis der Entscheidung bei Unsicherheit. 2. Auflage, IV, 125 Seiten 4°. 1969. DM 16,–

Vol. 2: U. N. Bhat, A Study of the Queueing Systems M/G/1 and GI/M/1. VIII, 78 pages. 4°. 1968. DM 16,–

Vol. 3: A. Strauss, An Introduction to Optimal Control Theory. VI, 153 pages. 4°. 1968. DM 16,–

Vol. 4: Branch and Bound: Eine Einführung. 2., geänderte Auflage. Herausgegeben von F. Weinberg. VII, 174 Seiten. 4°. 1972. DM 18,–

Vol. 5: Hyvärinen, Information Theory for Systems Engineers. VIII, 205 pages. 4°. 1968. DM 16,–

Vol. 6: H. P. Künzi, O. Müller, E. Nievergelt, Einführungskursus in die dynamische Programmierung. IV, 103 Seiten. 4°. 1968. DM 16,–

Vol. 7: W. Popp, Einführung in die Theorie der Lagerhaltung. VI, 173 Seiten. 4°. 1968. DM 16,–

Vol. 8: J. Teghem, J. Loris-Teghem, J. P. Lambotte, Modèles d'Attente M/G/1 et GI/M/1 à Arrivées et Services en Groupes. IV, 53 pages. 4°. 1969. DM 16,–

Vol. 9: E. Schultze, Einführung in die mathematischen Grundlagen der Informationstheorie. VI, 116 Seiten. 4°. 1969. DM 16,–

Vol. 10: D. Hochstädter, Stochastische Lagerhaltungsmodelle. VI, 269 Seiten. 4°. 1969. DM 18,–

Vol. 11/12: Mathematical Systems Theory and Economics. Edited by H. W. Kuhn and G. P. Szegö. VIII, IV, 486 pages. 4°. 1969. DM 34,–

Vol. 13: Heuristische Planungsmethoden. Herausgegeben von F. Weinberg und C. A. Zehnder. II, 93 Seiten. 4°. 1969. DM 16,–

Vol. 14: Computing Methods in Optimization Problems. Edited by A. V. Balakrishnan. V, 191 pages. 4°. 1969. DM 16,–

Vol. 15: Economic Models, Estimation and Risk Programming: Essays in Honor of Gerhard Tintner. Edited by K. A. Fox, G. V. L. Narasimham and J. K. Sengupta. VIII, 461 pages. 4°. 1969. DM 24,–

Vol. 16: H. P. Künzi und W. Oettli, Nichtlineare Optimierung: Neuere Verfahren, Bibliographie. IV, 180 Seiten. 4°. 1969. DM 16,–

Vol. 17: H. Bauer und K. Neumann, Berechnung optimaler Steuerungen, Maximumprinzip und dynamische Optimierung. VIII, 188 Seiten. 4°. 1969. DM 16,–

Vol. 18: M. Wolff, Optimale Instandhaltungspolitiken in einfachen Systemen. V, 143 Seiten. 4°. 1970. DM 16,–

Vol. 19: L. Hyvärinen, Mathematical Modeling for Industrial Processes. VI, 122 pages. 4°. 1970. DM 16,–

Vol. 20: G. Uebe, Optimale Fahrpläne. IX, 161 Seiten. 4°. 1970. DM 16,–

Vol. 21: Th. Liebling, Graphentheorie in Planungs- und Tourenproblemen am Beispiel des städtischen Straßendienstes. IX, 118 Seiten. 4°. 1970. DM 16,–

Vol. 22: W. Eichhorn, Theorie der homogenen Produktionsfunktion. VIII, 119 Seiten. 4°. 1970. DM 16,–

Vol. 23: A. Ghosal, Some Aspects of Queueing and Storage Systems. IV, 93 pages. 4°. 1970. DM 16,–

Vol. 24: Feichtinger, Lernprozesse in stochastischen Automaten. V, 66 Seiten. 4°. 1970. DM 16,–

Vol. 25: R. Henn und O. Opitz, Konsum- und Produktionstheorie. I. II, 124 Seiten. 4°. 1970. DM 16,–

Vol. 26: D. Hochstädter und G. Uebe, Ökonometrische Methoden. XII, 250 Seiten. 4°. 1970. DM 18,–

Vol. 27: I. H. Mufti, Computational Methods in Optimal Control Problems. IV, 45 pages. 4°. 1970. DM 16,–

Vol. 28: Theoretical Approaches to Non-Numerical Problem Solving. Edited by R. B. Banerji and M. D. Mesarovic. VI, 466 pages. 4°. 1970. DM 24,–

Vol. 29: S. E. Elmaghraby, Some Network Models in Management Science. III, 177 pages. 4°. 1970. DM 16,–

Vol. 30: H. Noltemeier, Sensitivitätsanalyse bei diskreten linearen Optimierungsproblemen. VI, 102 Seiten. 4°. 1970. DM 16,–

Vol. 31: M. Kühlmeyer, Die nichtzentrale t-Verteilung. II, 106 Seiten. 4°. 1970. DM 16,–

Vol. 32: F. Bartholomes und G. Hotz, Homomorphismen und Reduktionen linearer Sprachen. XII, 143 Seiten. 4°. 1970. DM 16,–

Vol. 33: K. Hinderer, Foundations of Non-stationary Dynamic Programming with Discrete Time Parameter. VI, 160 pages. 4°. 1970. DM 16,–

Vol. 34: H. Störmer, Semi-Markoff-Prozesse mit endlich vielen Zuständen. Theorie und Anwendungen. VII, 128 Seiten. 4°. 1970. DM 16,–

Vol. 35: F. Ferschl, Markovketten. VI, 168 Seiten. 4°. 1970. DM 16,–

Vol. 36: M. P. J. Magill, On a General Economic Theory of Motion. VI, 95 pages. 4°. 1970. DM 16,–

Vol. 37: H. Müller-Merbach, On Round-Off Errors in Linear Programming. VI, 48 pages. 4°. 1970. DM 16,–

Vol. 38: Statistische Methoden I, herausgegeben von E. Walter. VIII, 338 Seiten. 4°. 1970. DM 22,–

Vol. 39: Statistische Methoden II, herausgegeben von E. Walter. IV, 155 Seiten. 4°. 1970. DM 16,–

Vol. 40: H. Drygas, The Coordinate-Free Approach to Gauss-Markov Estimation. VIII, 113 pages. 4°. 1970. DM 16,–

Vol. 41: U. Ueing, Zwei Lösungsmethoden für nichtkonvexe Programmierungsprobleme. VI, 92 Seiten. 4°. 1971. DM 16,–

Vol. 42: A. V. Balakrishnan, Introduction to Optimization Theory in a Hilbert Space. IV, 153 pages. 4°. 1971. DM 16,–

Vol. 43: J. A. Morales, Bayesian Full Information Structural Analysis. VI, 154 pages. 4°. 1971. DM 16,–

Vol. 44: G. Feichtinger, Stochastische Modelle demographischer Prozesse. XIII, 404 Seiten. 4°. 1971. DM 28,–

Vol. 45: K. Wendler, Hauptaustauschschritte (Principal Pivoting). II, 64 Seiten. 4°. 1971. DM 16,–

Vol. 46: C. Boucher, Leçons sur la théorie des automates mathématiques. VIII, 193 pages. 4°. 1971. DM 18,–

Vol. 47: H. A. Nour Eldin, Optimierung linearer Regelsysteme mit quadratischer Zielfunktion. VIII, 163 Seiten. 4°. 1971. DM 16,–

Vol. 48: M. Constam, Fortran für Anfänger. VI, 143 Seiten. 4°. 1971. DM 16,–

Vol. 49: Ch. Schneeweiß, Regelungstechnische stochastische Optimierungsverfahren. XI, 254 Seiten. 4°. 1971. DM 22,–

Vol. 50: Unternehmensforschung Heute – Übersichtsvorträge der Züricher Tagung von SVOR und DGU, September 1970. Herausgegeben von M. Beckmann. VI, 133 Seiten. 4°. 1971. DM 16,–

Vol. 51: Digitale Simulation. Herausgegeben von K. Bauknecht und W. Nef. IV, 207 Seiten. 4°. 1971. DM 18,–

Vol. 52: Invariant Imbedding. Proceedings of the Summer Workshop on Invariant Imbedding Held at the University of Southern California, June – August 1970. Edited by R. E. Bellman and E. D. Denman. IV, 148 pages. 4°. 1971. DM 16,–

Vol. 53: J. Rosenmüller, Kooperative Spiele und Märkte. IV, 152 Seiten. 4°. 1971. DM 16,–

Vol. 54: C. C. von Weizsäcker, Steady State Capital Theory. III, 102 pages. 4°. 1971. DM 16,–

Vol. 55: P. A. V. B. Swamy, Statistical Inference in Random Coefficient Regression Models. VIII, 209 pages. 4°. 1971. DM 20,–

Vol. 56: Mohamed A. El-Hodiri, Constrained Extrema. Introduction to the Differentiable Case with Economic Applications. III, 130 pages. 4°. 1971. DM 16,–

Vol. 57: E. Freund, Zeitvariable Mehrgrößensysteme. VII, 160 Seiten. 4°. 1971. DM 18,–

Vol. 58: P. B. Hagelschuer, Theorie der linearen Dekomposition. VII, 191 Seiten. 4°. 1971. DM 18,–

Vol. 59: J. A. Hanson, Growth in Open Economics. IV, 127 pages. 4°. 1971. DM 16,–

Vol. 60: H. Hauptmann, Schätz- und Kontrolltheorie in stetigen dynamischen Wirtschaftsmodellen. V, 104 Seiten. 4°. 1971. DM 16,–

Vol. 61: K. H. F. Meyer, Wartesysteme mit variabler Bearbeitungsrate. VII, 314 Seiten. 4°. 1971. DM 24,–

Vol. 62: W. Krelle u. G. Gabisch unter Mitarbeit von J. Burgermeister, Wachstumstheorie. VII, 223 Seiten. 4°. 1972. DM 20,–

Vol. 63: J. Kohlas, Monte Carlo Simulation im Operations Research. VI, 162 Seiten. 4°. 1972. DM 16,–

Vol. 64: P. Gessner u. K. Spremann, Optimierung in Funktionenräumen. IV, 120 Seiten. 4°. 1972. DM 16,–

Vol. 65: W. Everling, Exercises in Computer Systems Analysis. VIII, 184 pages. 4°. 1972. DM 18,–

Vol. 66: F. Bauer, P. Garabedian and D. Korn, Supercritical Wing Sections. V, 211 pages. 4°. 1972. DM 20,–

Vol. 67: I. V. Girsanov, Lectures on Mathematical Theory of Extremum Problems. V, 136 pages. 4°. 1972. DM 16,–

Vol. 68: J. Loeckx, Computability and Decidability. An Introduction for Students of Computer Science. VI, 76 pages. 4°. 1972. DM 16,–

Vol. 69: S. Ashour, Sequencing Theory. V, 133 pages. 4°. 1972. DM 16,–

Vol. 70: J. P. Brown, The Economic Effects of Floods. Investigations of a Stochastic Model of Rational Investment Behavior in the Face of Floods. V, 87 pages. 4°. 1972. DM 16,–

Vol. 71: R. Henn und O. Opitz, Konsum- und Produktionstheorie II. V, 134 Seiten. 4°. 1972. DM 16,–

Vol. 72: T. P. Bagchi and J. G. C. Templeton, Numerical Methods in Markov Chains and Bulk Queues. XI, 89 pages. 4°. 1972. DM 16,–

Vol. 73: H. Kiendl, Suboptimale Regler mit abschnittweise linearer Struktur. VI, 146 Seiten. 4°. 1972. DM 16,–

Vol. 74: F. Pokropp, Aggregation von Produktionsfunktionen. VI, 107 Seiten. 4°. 1972. DM 16,–

Vol. 75: GI-Gesellschaft für Informatik e. V. Bericht Nr. 3. 1. Fachtagung über Programmiersprachen · München, 9–11. März 1971. Herausgegeben im Auftrag der Gesellschaft für Informatik von H. Langmaack und M. Paul. VII, 280 Seiten. 4°. 1972. DM 24,–

Vol. 76: G. Fandel, Optimale Entscheidung bei mehrfacher Zielsetzung. 121 Seiten. 4°. 1972. DM 16,–

Vol. 77: A. Auslender, Problemes de Minimax via l'Analyse Convexe et les Inégalités Variationnelles: Théorie et Algorithmes. VII, 132 pages. 4°. 1972. DM 16,–

Vol. 78 : GI-Gesellschaft für Informatik e.V. 2. Jahrestagung, Karlsruhe, 2.–4. Oktober 1972. Herausgegeben im Auftrag der Gesellschaft für Informatik von P. Deussen. XI, 576 Seiten. 4°. 1973. DM 36,–

Vol. 79 : A. Berman, Cones, Matrices and Mathematical Programming. V, 96 pages. 4°. 1973. DM 16,–

Vol. 80: International Seminar on Trends in Mathematical Modelling, Venice, 13–18 December 1971. Edited by N. Hawkes. VI, 288 pages. 4°. 1973. DM 24,–

Vol. 81: Advanced Course of Software Engineering. Herausgegeben von F. L. Bauer. XII, 545 pages. 4°. 1973. DM 32,–

Vol. 82: R. Saeks, Resolution Space, Operators and Systems. X, 267 pages. 4°. 1973. DM 22,–